Pursuant to article 1 of the Convention signed in Paris on 14th December, 1960, and which came into force on 30th September, 1961, the Organisation for Economic Co-operation and Development (OECD) shall promote policies designed:

- to achieve the highest sustainable economic growth and employment and a rising standard of living in Member countries, while maintaining financial stability, and thus to contribute to the development of the world economy;
- to contribute to sound economic expansion in Member as well as non-member countries in the process of economic development; and
- to contribute to the expansion of world trade on a multilateral, non-discriminatory basis in accordance with international obligations.

The original Member countries of the OECD are Austria, Belgium, Canada, Denmark, France, the Federal Republic of Germany, Greece, Iceland, Ireland, Italy, Luxembourg, the Netherlands, Norway, Portugal, Spain, Sweden, Switzerland, Turkey, the United Kingdom and the United States. The following countries acceded subsequently through accession at the dates hereafter: Japan (28th April, 1964), Finland (28th January, 1969), Australia (7th June, 1971) and New Zealand (29th May, 1973).

The Socialist Federal Republic of Yugoslavia takes part in some of the work of the OECD (agreement of 28th October, 1961).

Publié en français sous le titre:

TARIFICATION
DES SERVICES
RELATIFS A L'EAU

This report has been prepared as part of the programme of the OECD's Environment Committee and its preparation has been overseen by the Steering Group on the Economic Aspects of Water Conservation consisting of representatives from Member countries assisted by Dr. F. Juhasz from the OECD Secretariat.

The report was prepared and written by Mr. P. Herrington, Department of Economics of the University of Leicester, United Kingdom, and is made available to the public on the responsibility of the Secretary-General of OECD.

Also available

ENVIRONMENTAL POLICY AND TECHNICAL CHANGE (September 1985)
(97 85 07 1) ISBN 92-64-12731-3   104 pages   £9.00   US$18.00   F90.00   DM40.00

THE MACRO-ECONOMIC IMPACT OF ENVIRONMENTAL EXPENDITURE (August 1985)
(97 85 06 1) ISBN 92-64-12716-X   120 pages   £7.50   US$15.00   F75.00   DM33.00

OECD ENVIRONMENTAL DATA - COMPENDIUM 1985 (June 1985) bilingual
(97 85 05 3) ISBN 92-64-02678-9   298 pages   £15.00   US$30.00   F150.00   DM67.00

THE STATE OF THE ENVIRONMENT - 1985 (June 1985)
(97 85 04 1) ISBN 92-64-12713-5   272 pages   £16.50   US$33.00   F165.00   DM73.00

ENVIRONMENT AND ECONOMICS. Results of the International Conference on Environment and Economics, 18th-21st June, 1984 (April 1985)
(97 85 01 1) ISBN 92-64-12691-0   248 pages   £12.00   US$24.00   F120.00   DM53.00

Prices charged at the OECD Bookshop.

THE OECD CATALOGUE OF PUBLICATIONS and supplements will be sent free of charge on request addressed either to OECD Publications Service, Sales and Distribution Division, 2, rue André-Pascal, 75775 PARIS CEDEX 16, or to the OECD Sales Agent in your country.

TABLE OF CONTENTS

SUMMARY AND CONCLUSION

1. BACKGROUND

As part of its work programme on natural resource management, the OECD's Environment Committee carried out an examination of pricing of water and related services.

In 1983 the Environment Committee formed a Steering Group on the Economics Aspects of Water Conservation. The overall reason for the involvement of the Environment Committee in this area was the concern to ensure that the water resources of OECD countries were used in the most economically and environmentally efficient manner. For both economic and environmental reasons it has become increasingly difficult in many areas of the OECD to meet water demands in their various forms and an investigation of, and emphasis on, water demand management was considered warranted.

The mandate of the Group was to:

i)   Assess, in the context of prevailing conditions in Member countries, the contribution of economic techniques, in particular water pricing, for developing practical options for the efficient management of demand and supply of the appropriate quality of water;

ii) Develop guidelines for promoting the conservation, reallocation and reuse of water resources through the combination of regulatory and economic instruments; and

iii) Assess, to the extent possible the relevance of the proposed guidelines for the conservation of international water resources.

The project was limited in both the time and resources available and it was agreed to concentrate on those issues within the subject area that were considered to be of highest priority by Member countries. Following a survey of member country requirements the role of water pricing and the forecasting of water demand were selected as two areas for work. The role of water demand forecasting is covered in another paper "Water Demand Forecasting in OECD countries".

For the purposes of the report, water supply and disposal services were defined to include potable and non-potable piped water supplies, direct withdrawals of water from its sources, and disposal and treatment of effluents through a sewerage system or directly to receiving waters.

9

2. THE RESULTS

Pricing Practices in Member Countries

Water pricing policies are many and varied across OECD Member countries but most remain based on financial rather than economic considerations. Such a financial approach attempts to ensure that total revenue earned balances total costs. The primacy of the financial requirement has meant that in most countries, the water industry has been preoccupied with price/rate levels and until recently little concerned with price/rate structures. Pricing systems recently in use for piped services (water supply, sewerage and sewage treatment), were classified as follows:

-- Flat rate tariffs (where water fees are not directly related to quantities of water used). It was found that those charges were levied on many bases including the number of residents, number and type of rooms, number of taps, size of inflow pipe, ground area or property value.

-- Average cost pricing (where all water services costs, other than access costs, are grouped together and divided among the total number of units expected to be sold, to generate a unit cost).

-- Declining block tariffs (where succeeding blocks of units of water consumed are sold at lower and lower prices. A fixed or minimum billing charge is usually included).

-- Increasing or progressive block tariffs (where succeeding blocks of units of water are sold at higher and higher prices). It was found that these tariffs were becoming increasingly common, and reflected at least initially, income redistribution objectives (provision of an initial, basic supply at a low rate, with additional consumption becoming progressively more expensive). It was also found that the system was used to promote efficiency and conservation objectives, although it is noted that the evidence of the effectiveness of such tariffs in the industrial sector is not clear, given the concurrent economic recession and restructuring away from large water using industries.

-- Two-part tariffs (including a fixed element, often varying according to some characteristic of the user, and average cost pricing in the form of a single volumetric charge).

For 'natural' services (effluent discharge and direct abstractions, including irrigation demands), charging practices also vary greatly:

-- license fees (sometimes of only nominal amounts);

-- other flat-rate charges;

-- charges related to authorised abstractions or discharges;

-- charges related to actual abstractions or discharges.

10

For sewage and sewerage disposal (S&SD) services it was found that because of the great expense and difficulty of separately measuring waste volumes for residential users of sewers and sewage treatment, S&SD charges were normally hinged on the structure of the public water supply tariff. For domestic consumers, the input of water was found to be a satisfactory proxy for the volume of sewage generated. Thus, it was found that the effects of pricing S&SD in different ways would be hard to isolate from water supply charges, since the two were often presented together to the customer as a combined tariff. For industrial users of public sewers, the characteristics of water borne wastes differ enormously from one discharger to another and in this situation a charging system reflecting differentiated charges for specialized treatment of different characteristics of wastes may well be justified. The evidence suggested that price increases cause significant reductions in the demand for trade effluent treatment services.

For other 'natural' services it was found that different types of legal environment (private water rights, government ownership, and common or collective ownership), influenced efficient allocation, the development of water services and the role of charging policy.

Two contrasting traditions of private property in water were apparent in a number of countries: the riparian doctrine, linking ownership of water rights to ownership of the adjacent, or overlying land, and the doctrine of appropriation, when the water right is acquired by use.

Much variation in the control of direct abstractions was apparent among members states. In most countries, licenses were required for both surface and ground water abstractions, facilitating some regulation of demands. Charges over and above the often nominal license fees were less common than permits, and may take the form of annual charges and unit quantity charges. Much evidence exists that the charging schemes described were and still are designed to generate revenue to pay for water control and supply works necessary for the abstractions.

For irrigation services it was found that, among OECD Member states, the importance of irrigation varied enormously. Evidence suggested that the demand for irrigation water was highly responsive to price changes. The presumption must be made that significant price elasticities would be confirmed if the research effort could be mounted. Charging systems for irrigation water were found to fit into one or both of two categories: flat-rate or fixed charges associated with a "normal" allocation of water and unrelated to the volume actually abstracted, and charges related to authorized, actual or "excess" volumes.

Looking across the range of all the above mentioned water services, it was found that the evidence concerning the effects of these different tariff structures was consistent. Flat-rate tariffs have generally meant higher use of the water service and therefore over-building of supply systems and over-utilisation of the natural resource. Highly-subsidised (and therefore very low) volumetric charges may produce the same result, especially in the case of agricultural irrigation. The introduction of volumetric charging systems and of seasonal tariffs, and the reduction of very large 'free allowance' tranches in tariff structures, have all been found to induce significant reductions in demands and consequently economic and environmental benefits.

Metering

It was also found that the role and impact of metering water use was very important. There is an enormous variation in the extent of domestic metering in OECD countries.

The metering decision has two dimensions, equity and efficiency. Metering was found to be equitable in that it permitted volumetric charging, which meant payment for the quantity of water used (and, approximately, the quantity disposed of into the sewers).

In assessing efficiency, a fundamental distinction must be made between economic and financial appraisal. Cost-benefit analysis is recommended to determine the desirability on efficiency grounds of changing water consumers over to a metering and volumetric charging system. The financial appraisal approach assesses the balance in the changes in cash inflows and outflows that would result from metering, and would be necessary for calculating the consequences for consumers' charges.

Subsidisation

In relation to subsidies in water services, it was found that the extent of subsidisation, in the form of grants or low-interest loans by governments to water utilities, varied greatly within and across OECD countries, being generally higher in OECD's "off-centre" members in North American, Japan and Australia.

With few exceptions, current expenditures on public water supply systems (operations and maintenance and debt interest) are covered by user charges. Subsidisation of capital expenditures is much more frequent, and also greater. In waste water services, subsidies are in general higher. The highest subsidies were however found for irrigation projects.

3. WATER PRICING POLICIES

Based on this report the Environment Committee concluded that ideally water should be provided and allocated in an economically efficient manner so that the net benefits that the community as a whole derived from the use of water resources were maximised.

A charging system that is efficient in this sense is known as a marginal cost pricing system. In such a system the price reflects the incremental costs to the community of satisfying marginal demands. These incremental costs include quantitative and qualitative resource depletion costs, damage costs and the various capital and operating resource costs. Thus the price of the last litre of water used or disposed of would be equal to the true marginal cost of 'providing' that water service, be it 'natural' or 'man-made'. In this way, the marginal value of the litre abstracted, consumed or disposed of in each use would be the same; the supply system would be used at its economic optimum rate; and, in the long-run, the supply system would be constructed and maintained at the optimal economic scale. In every sense, then, an economically efficient use of resources would be promoted.

It was established that <u>inefficient use</u> of water resources could have <u>significant environmental impacts</u>. Artificially low water prices and other policies could encourage overuse of water which could lead, for example, to the construction of unnecessary water storage facilities or to the excessive drawdown of underground aquifers which may have major impacts upon the aquatic environment. Also, low water prices could lead to excessive use of water for irrigation purposes which could result in increased nitrate, phosphate and pesticide contamination of aquifers and increased soil degradation through compaction and salinisation. Inefficient water pricing in industrial applications could also lead to the excessive use of water for diluting effluents to meet legal concentration standards. This may also increase the difficulty of eventual pollutant removal.

The diversity of the water market and the need for pricing policies to be adaptable and applicable to the different characteristics of the agricultural, domestic, commercial and other (recreational, environmental) sectors of the market was noted. The differences between public and private markets for water provision was also recognised.

It was recognised, however, that the water market was not a "perfect market" in the theoretical sense and that, in the real world, deviations from the principle of marginal cost pricing would necessarily occur. It was recognised, nevertheless, that from the viewpoint of economic efficiency and environmental quality, a system of pricing, based on the principle of marginal cost pricing, such as the increasing block tariff, would be preferred.

At the same time there was a realisation that the use of water resources for the provision of water services sometimes imposed costs on third parties and generated benefits to those other than the direct consumers of the services (for example, public health benefits can accrue to neighbours from a household's use of a piped sewage disposal service; loss of ecological value may occur when a water storage facility interferes with a natural stream). Such costs should be taken account of in the measurement of marginal costs, but the existence of external benefits would in principle alter the marginal cost pricing basis. This could cause practical difficulties, so that a marginal cost pricing system composed of a two-part tariff (a fixed element and a single volumetric rate) is generally to be preferred. For direct abstractions, the marginal capital cost of maintaining supplies at a given quality should provide the basis for charges for <u>authorised</u> abstractions, while <u>actual</u> abstractions should reflect the marginal operating cost of the supply system (including any damage and resource depletion costs).

Thus, efficient pricing methods, based on the principles of marginal cost pricing but responsive to the day to day realities of water resources management, were recognised as being an important policy instrument that could improve the economic and environmental efficiency of water resources use.

In addition, it was recognised that pricing policies could be used to best effect as the major component of a "package" of measures containing demand management instruments. It was also recognised that a precondition for an effective pricing policy was the installation of metering devices. A decision whether or not to introduce a more efficient pricing scheme would depend on the costs and benefits of metering. Current rapid advances in the technology of sophisticated metering were noted.

Other Influences on Pricing Policies

The investigation established that, in order to be effective and practical, pricing policies should be responsive to the following factors and should be combined in a complementary package of demand management measures:

i) Equity or fairness

This was recognised as a difficult objective to define. There were found to be a number of concepts of equity. They ranged from broader notions, including 'social pricing' whereby no consumer should be prevented by income considerations from enjoying the benefits of water, to narrower concepts such as that provided by the requirement that each consumer should pay the same per unit of water service received regardless of its cost.

The validity of alternative approaches to the broader consideration of equity was recognised. In some countries, water pricing is structured to foster development in agricultural and industrial sectors, to lighten charges on isolated communities or to help low-income households; in others the view is taken that it is preferable to adopt an economically efficient water pricing system in combination with a social security system for those consumers disadvantaged by the policy. It was thought that any financial subsidies to the water services as a whole, or to one group of water service users from another, should be made explicit and be justified by arguments for special treatment.

ii) Financial Requirements

Whilst water management authorities were not usually required to cover all economic costs of their operations (resources, capital and operating), they were often required to cover a substantial proportion of their financial costs (usually operating costs and at least a proportion of their capital costs). It was found that it was possible to pursue both financial and efficiency objectives and that the problems of designing a pricing system to meet efficiency and financial objectives were not insurmountable.

iii) Consumer Acceptability

It was found that any pricing system must be easily understood and acceptable to the consumers of water pricing services. This naturally ruled out some very complex pricing systems which, although they remain faithful to the marginal cost pricing principle, may confuse and may therefore transmit incorrect signals to consumers. Nevertheless it was noted that this consideration did not preclude the use of a simple tariff structure based on efficiency principles.

iv) Administrative Costs

It was noted that any pricing system adopted should not impose large administrative costs. This is particularly the case in marginal cost pricing where the efficiency gains could be offset by administrative costs. This objective imposes effective limits on the feasible complexity and differentiation in charging systems used for allocative efficiency but again did not rule out the use of simple pricing systems based on efficiency principles.

v) Environmental Considerations

In addition it was noted that pricing systems should, where possible, reflect the ecological, recreational and environmental uses of water. It was observed that the more visible commercial and urban uses of water had traditionally taken precedence over these environmental and aesthetic uses. A pricing system should promote the sensible use of the environment and thus reflect the complete social and environmental costs of providing a water service. This would include consideration of the opportunity costs of use of the water resources base and of the longer-term aspects of water resources use.

At the same time it was recognised that the pursuit of these objectives was completely consistent with the use of simple pricing systems based on principles of economic efficiency.

vi) Other Government Policies

As with the development of all policies, it was noted that it is necessary to pay due regard to other government policies (perhaps general economic, agriculture, labour or energy policies) to minimise the occurrence of conflicting policies. It was recognised that different policy instruments had different effects on the achievement of policy goals and that it was the responsibility of governments to ensure compatibility of both instruments and goals.

It was clear that these other objectives confronting water managers could be accommodated within the pursuit of the economic efficiency objective and that these other objectives need not necessarily be reasons for the adoption of pricing principles in contravention of economic efficiency principles.

4. CONCLUSIONS

Marginal Cost Pricing

The evidence gathered during the project on the pricing of water led to the conclusion that in order to ensure the most efficient use of a scarce natural resource, water management authorities in Member countries should consider the use of economically efficient pricing mechanisms, in all water uses, based on the objective of marginal cost pricing, as part of their overall approach to water demand management.

Such mechanisms, if introduced and administered correctly, should lead to more efficient use of the water resource. It was noted that given the many objectives confronting water resource managers, the strict adherence to the marginal cost pricing principle could create problems in meeting these objectives. Sensible use, in line with the principles of marginal cost pricing, would be likely to improve the management of this scarce resource. It was noted that in the absence of metering (for domestic, industrial and agricultural uses) only a flat rate pricing system could be used which would act as a disincentive to efficient water use.

The User Pays Principle

The findings of the report support the general applicability of the User-Pays Principle recognising at the same time the historical and social background of Member countries. The essential element in the User-Pays Principle is that an incentive is provided for the user to economise in the use of the service or natural resource. Users of the service would in aggregate pay the full cost of the provision of the service and a quantity and quality-based charging system would share out these costs among users. This is similar to, and indeed embraces, the more familiar 'Polluter-Pays Principle', where the polluter pays for the cost his pollution imposes on society. Under the User-Pays Principle, subsidisation by taxpayers and cross-subsidisation among water service users would be abolished unless special 'social' reasons existed for their retention. Ideally, marginal cost pricing should be applied under the User-Pays Principle. However, during the phasing-in period and as recognised above, other considerations might limit the full implementation of marginal cost pricing.

5. THE POLICY IMPLICATIONS

On the basis of the observations made, conclusions drawn and taking into account the historical and social background of Member countries, the Environment Committee has arrived at the following policy implications in relation to measures to be adopted by Member countries:

i) In the light of growing prospective demand-supply imbalances, water authorities in Member countries should pay increasing attention to the metering and pricing of all water services, flexibility in the use and reallocation of water rights, and other methods of managing demand.

ii) OECD countries should now consider the adoption of the User-Pays Principle for water and water related services, and in the application of this principle marginal cost pricing should be used. Under the User-Pays Principle the polluter remains responsible for the cost of pollution and prevention control, as agreed to by OECD

16

countries under the Polluter-Pays Principle. Further work should be undertaken to assure that these two principles are applied in an economically efficient and co-ordinated manner.

iii) The User-Pays Principle requires water authorities to take into account economic efficiency in the formulation of pricing structures and determination of tariff levels. They should also take account of environmental and conservation objectives, equity considerations and the need to raise sufficient revenue to cover costs.

iv) Financial subsidies to the water services as a whole, or to one group of water service users from another, should be justified by explicit arguments for the special treatment of the particular group.

Chapter I

THE WATER SERVICES: DEFINITION AND CONTEXT

1. DEFINITION

The water services to be examined cover the provision and disposal of water supplies. They may thus be listed as:

i) The provision of potable piped water supplies;
ii) The provision of non-potable piped water supplies;
iii) Direct abstractions (or withdrawals) which actually remove water from its source, however temporarily;
iv) The disposal and any subsequent treatment of effluents through a sewerage system;
v) Direct discharges of effluents to receiving waters.

(i) and (ii) are normally linked together under the umbrella title, 'the public water supply system'. Non-potable piped supplies are generally the junior partners in such systems, and in many regions are not provided at all.

All other water-related services including surface water drainage from property and highways will be omitted from consideration. Most of them relate to in situ rather than flow uses of water, the more important being flood prevention and protection, land drainage, recreation, and navigation.

2. ECONOMIC CONTEXT

To make water available how, when and where consumers demand it, economic resources (human, capital, knowledge) generally have to be applied to the natural resource. Water is thus granted the status of an economic good.

As is shown in Annex 1, the opportunity costs of the provision of a water supply or effluent disposal service, natural or fabricated, may be generally defined as:

Opportunity Costs = Resource-use Costs (i.e., the goods or services forgone by the commitment of economic resources in the operation and maintenance of the supply system)

+ Natural Resource Depletion Costs (qualitative and/or quantitative, reflecting the consequences of present use for future supplies)

+ Damage Costs (external effects).

A related definition of the opportunity costs of finance committed to capital or current expenditures on the provision of water services would draw attention to the <u>financial</u> context in which water authorities and undertakings have to operate. A more limited framework would thereby emerge, since financial compensation may be a poor indicator of resource-use costs and will often give no guide at all to natural resource depletion and damage costs. In that sense economic and financial appraisal of water services projects may at times give conflicting results. An overall resource conservation objective means, however, that economic appraisal should be considered in principle superior.

3. THE ECONOMIC QUESTIONS TO BE ANSWERED

All economic systems, including those designed to manage water resources and the derived water services, must answer six fundamental questions:

i) <u>What</u> goods and services should be produced?
ii) What <u>quality</u> of these goods and services should be produced?
iii) What <u>quantity</u> should be produced?
iv) <u>How</u> should they be produced?
v) <u>For whom</u> should they be produced?
vi) What <u>spare capacity</u>, if any, should be built into the supply system?

Every one of these questions reduces to one or more problems concerning the allocation of resources (either between the water industry and other economic sectors or within the water industry) or the distribution of water services. Such issues comprise the heartland of economics. How should they be dealt with?

4. THE NATURAL PROVISION OF WATER SERVICES

First the natural capacity of bodies or flows of water to provide supply or disposal services will be discussed. Three main approaches to the questions raised above may be identified.

First, reliance upon <u>the law</u>. So long as the rights of water users are clearly defined, it would in principle be possible to let the law and 'the market' alone be responsible for allocating the available resources. Conflicts of interest like those arising from external effects on other parties could be left to the courts to settle. Damages and compensation would be paid and no separate 'pricing system' would be required. If the law permitted the sale of rights, capital values, representing the present value of the expected future net benefits generated by ownership of the rights, would be established through the forces of supply and demand interacting in a market. Under certain assumptions which are rather unlikely to be encountered in practice, these market forces may tend to produce an optimal allocation of water services.

A second possibility is the erection of an <u>administrative system of rights to abstract and discharge</u>, incorporated in licences or permits which define closely how, when, where and in what quantities and qualities the service may be enjoyed. In principle the administrative agency could continue to seek the optimal allocation of the service concerned as demands and technologies change, by revocation, amendment and reallocation of rights. An important trade-off immediately comes into view, for the shorter the time-horizon attached to the conditions of the permit the more difficult it would be for an abstractor or discharger to plan its activities ahead. Such uncertainty might seriously inhibit some productive enterprise.

Finally, resort may be had to a <u>formal charging scheme</u>, expressly designed <u>to promote the optimal use of the abstraction or assimilation services</u> afforded by rivers or aquifers. As it is shown see in Chapters V to VII, below, an increasing number of countries have begun to move to such charging systems in recent years. One particular advantage is that they are able to generate more useful information for assessing the desirability of new works which would extend the capacity of a supply or assimilative network. Further discussion of these matters is deferred to Chapters V to VII.

5. THE FABRICATED PROVISION OF WATER SERVICES

In the case of man-made capacity to supply or to dispose of water, the major economic problem is the continuing need to reconcile limited supplies with what are normally increasing demands. Two broad approaches to the solution of this problem may be identified. First, the organisation of the piped services may be geared largely if not exclusively to supply-management. Demand-supply gaps will tend to be closed through supply-expansion or more effective use of existing supplies (e.g., through conjunctive use of surface and ground water), with the pricing/allocation system driven by equity considerations and the need to generate revenue. This approach characterised the activities of most water utilities in developed countries until recent years.

Alternatively, water authorities may consciously engage in demand management, seeking to influence the demands for water services such that what are deemed to be the 'reasonable' demands of consumers are satisfied. 'Reasonable' would normally be defined with respect to (a) the costs of supplying and disposing of water and (b) considerations of public health.

It is useful to specify four distinct types of demand-management policies: regulatory, tariff, educational and operational control.

<u>Regulatory</u> policies may be instigated by governments or water

authorities at a national, regional or local level. They find expression through the provision of regulations, laws, bye-laws and codes, concerning (i) the physical quantity of water used in the operation of water-using technology (e.g., toilets, showers, washing machines, irrigation systems) and (ii) emission standards which stipulate maximum strengths and quantities of liquid wastes.

Tariff policies for influencing demands have an obvious attraction: they can be the means whereby the consumer -- householder or enterprise -- may be confronted with the costliness of maintaining or extending the provision of the service. Where no significant externalities occur, there are evident advantages in the consumer being the judge of whether the service is desired and, if so, the quantities and qualities that are to be provided.

Prices can therefore serve as important policy instruments in answering the questions of Chapter I.3. Indeed, because the water industry is almost inevitably a monopoly whose activities are not regulated through competition of price and service as are those of most commercial enterprises and a number of other public utilities (e.g., gas, electricity and rail, in some countries), more than usual effort may have to be devoted to the development of charging policy to ensure that services are not operated or extended at a cost to the community that exceeds their value to users.

Education is sometimes seen as a demand-management policy in its own right, whereby a continuing stream of expenditures on information provision or moral suasion may be 'justified' by either the cost savings associated with a more economical use of water or the improved allocation of limited supplies thereby facilitated. Alternatively education can be viewed as an essential complement to the regulatory and tariff policies just described. Its role then lies in increasing knowledge of, and explaining, an authority's regulations and tariffs. For example, it would be useful to publicise a listing of the different appliances on the market which meet a revised plumbing code; and information could be made available to metered householders giving examples of potential water-economising measures and the magnitude of the financial gains to a consumer which are likely to ensue.

Operational control policies cover leakage and pressure control (water supply) and infiltration (into sewers). Leakage and infiltration control policies should be designed such that the net social benefits are maximised. Detection and prevention of leakage and infiltration should be taken to the point where the extra costs of further action just begin to outweigh the extra benefits (primarily, the deferment of new supplies and savings in operating costs). Variable pressure control results in a reduction in bursts as well as a more equitable delivery of the standard of service. Electronic telemetry offers enormous scope for the development of automatic pressure control.

To investigate all these varieties of demand-management would be to travel way beyond the remit of this study. Succeeding sections of the text will therefore be devoted to an examination of actual and desirable charging policies for the water industry.

6. LIMITATIONS OF PRICING

It should be realised, however, that there are certain dangers in asking for too much from a pricing system. The first point to recognise is that in practice elasticities of demand for the water services may be so low that sophisticated charging mechanisms are simply not worthwhile. That is, the costs of establishing, operating and updating complex tariff structures may simply be too high. This argument is developed in a number of different ways in Chapters II.6, II.7, III.5 and VIII, below.

Second, the more complex is a charging scheme the higher the proportion of consumers who are unlikely to be able to understand it fully and therefore are unable to react 'rationally' to it (see II.6, below). Third, consumers may well possess imperfect technical and financial information concerning the technology available for economising on the use of water or for reducing the volume (or strength) of generated effluents. Clearly this will impose a further limit on the reaction that householders and enterprises can make to charges.

Fourth, capital market imperfections may result in a significant bias against investment expenditure (e.g., on recirculation plant or new treatment technology) and in favour of the payment of increased operating expenditures (in the form of increased water supply or effluent charges). This can affect firms, in which current expenditures are seldom subject to as much scrutiny as investment expenditures (a local branch or plant may have to refer the latter to head office, but not normally the former), and also householders who might, for example, be inhibited from taking up a metering option if capital and installation costs have to be financed with a personal loan. Fifth, most pricing systems for commercial enterprises are designed on the assumption that firms are attempting to maximise profits. Alternative commercial objectives such as revenue or sales maximisation may mean that actual responses to charging systems differ from those desired.

Finally, the limitation of prices in helping to establish the correct qualitative dimensions of water services should be appreciated. Whereas the design of an effluent disposal charging scheme may include quantitative and qualitative dimensions, and thus encourage or discourage pre-discharge treatment by firms (hence affecting the quality of the received effluent), it is difficult to imagine pricing being used to help consumers choose among a number of qualities of water supply (save in the trivial sense that a water authority having decided on, say, a dual-quality supply system, the careful monitoring of demands as prices change may generate information useful for quality amendments over time). Thus the 'automatic' management of supply quantity, and effluent quantity and quality, which may -- at least in principle -- be achieved by suitable design of a charging scheme, is not practically feasible in the case of supply quality. Similarly a pricing system is of only indirect help in decisions concerning the establishment of optimal standards of service (e.g., reliability).

Chapter II

PRINCIPLES OF WATER PRICING

It will be useful to set out a formal listing of the criteria which should or may inform the design of sensible charging systems for water services.

An economic approach to water conservation suggests that efficient allocation of resources should be the prime objective of a charging system. The other criteria to be listed act largely as constraints which generate pressures on tariff designers. Some of the criteria may be mutually inconsistent such that significant trade-offs arise. Thus in practice tariff structures are bound to reflect compromise. The experience of Member countries as reflected in this report makes it seem certain, however, that long-term water conservation requires first and foremost the use of prices as incentives to further the rational use and allocation of water services.

1. ALLOCATIVE EFFICIENCY

Allocative efficiency means that water services should be provided such that the community's net benefits are maximised. Ideally, this would determine both quantity and quality, and, where the water service is priced, it implies that the price should reflect the incremental costs to the community of satisfying marginal demands. Such a charging system is usually known as marginal cost pricing.

The provision of water supply or disposal services raises the same questions of allocative efficiency as the supply of any other service produced by the private or public sectors. There are three situations to be considered.

Suppose first that the quantity of a service available is fixed. Efficient allocation and use then requires that the marginal value to the community from the last litre abstracted, consumed or disposed of by each user or user group and in each use should be the same. If the marginal values are not equal, then the community's welfare could be augmented by reassigning some of the service to the user or usage which offers the greater net income, the 'better opportunity' or the 'higher level of satisfaction' (assuming the last two terms can be satisfactorily defined). For even if some consumers were to be harmed by such reallocation, in principle the greater total income generated may be redistributed such that no consumer is worse off and some consumers are better off. It may be easily appreciated how inflexible and often historically-rooted systems of water rights, combined with the absence of markets in those rights or the services that flow from them, produces situations where water is very inefficiently allocated. For example, private agricultural interests with large and long-established water entitlements may be able to take their abstraction of river water to the point where the marginal value is much below that of a public water utility with a more recent and perhaps comparatively limited entitlement. In such a situation, reallocation of water use from agriculture to the public water supply might increase overall welfare dramatically.

Second, consider the decision to increase output from a given supply system. The extra operating resources needed (e.g., chemicals for purification, energy for pumping) will be unavailable for use in other sectors of the economy. Such a resource switch is economically sensible so long as the benefits of the extra output outweigh the opportunity costs of its provision.

Similarly -- and finally -- the decision to maintain or expand a supply system raises questions concerning the allocation of the economy's capital resources. To achieve the 'correct' balance between investment in water services and investment in other sectors of the economy full account must be taken of both the capital and the complementary operating costs, and also any damage and resource depletion costs, arising from the water services investment. Again, the benefits must be judged at least to outweigh the extra costs for the investment to be justified.

The last three paragraphs may be summarised in a slightly different language. Singleminded pursuit of the allocative efficiency objective would dictate that water services should be allocated and provided, both in the short-run and in the long-run, in the quantities and (where feasible) of the qualities which maximise the net benefits of provision (the excess of benefits over all costs).

Such an efficient use of resources is promoted if prices carry the correct messages to users; that is, if prices reflect the costs to the community of satisfying marginal demands. This is the principle of MARGINAL COST PRICING. It means that at the margin, i.e., for the last unit consumed, prices should reflect the incremental costs imposed on the economy in meeting those demands or the savings which could be made if they are not met.

If we were able to 'isolate' the transactions in water services from the rest of the economy, in the sense of being able to assume that the actual quantities of water services consumed had no significant effects on the demands for other goods and services (and vice-versa), then the marginal cost pricing requirement could be made more precise: that prices should equal marginal costs, so long as by 'marginal costs' we mean marginal opportunity costs. In the absence of such an assumption, i.e., when the water industry is interdependent with a 'deviant' sector like a monopoly, then the equating of price and marginal cost is in theory unjustified and 'optimal' departures from price/marginal cost equality may be calculated only when precise (and quantitative) information is available concerning the nature of the independence, e.g., in the form of the relevant cross-elasticities of demand (Rees, 1984). However, shortage of information will usually mean that the relationship of price to marginal cost required to satisfy efficiency criteria in the water services is assumed to be one of equality.

A second 'correction' to the marginal-cost-based price would be necessary if the benefits arising from the extra provision of a water service to an individual consumer were not restricted to that consumer, e.g., the public health benefits accruing to neighbours from a household's use of the piped sewage disposal service. In this situation the individual consumer's preparedness-to-pay, probably the best estimate available of the value of the marginal service to him or her, will understate the benefit to the community so long as the extra ('external') benefits are not all concentrated on non-marginal units of consumption. Only if the marginal opportunity of provision of the service were constant would the optimal price be left in the long-run unaffected (for then the amount of the service consumed would not affect the marginal cost of its provision). More fundamental questions about the optimal financing of the provision of public goods are opened up by these considerations, and these will be briefly revisited in IV.1, below.

Further concern with the exclusive use of marginal cost pricing to solve allocation problems arises in the situation where supply is unusually limited (e.g., for water supply, in a drought). At such times prices could in principle be raised until demands equal supplies, so that there would be true 'rationing by price', clearly favouring those with the most ability to pay. A community may understandably find such total reliance on the pricing system unacceptable because of the resulting hardship for poorer consumers.

We also note here that the water industry is exceptional in the amount of capital it has tied up to cope with peak demands. Climatic variation is responsible for much unpredictable change in demands, especially in the domestic and agricultural sectors, and other social factors make for diurnal (within the day), weekly and longer-term demand variation. On the supply side, climatic factors determine, often with a long and distributed lag, the quantities of water available for abstraction and consumption within a given supply system. In principle, therefore, the objective of allocative efficiency should give rise to tariff design which, responding to such supply and demand variation, will differentiate prices by time of day, day of the week, season of the year, degree of water shortage and also by the extent of simultaneous demand by other customers.

Pricing policy must also have regard to a longer time dimension. If increasing scarcity is anticipated, then the correct social policy may be to make the resource or service expensive straightaway. The theory of optimal resource use over time recognises the importance of present price signals to consumers in encouraging conservation (thus deferring the time of real scarcity) and promoting sensible decisions now about perhaps long-life investments complementary to or substitutes for the use of water services. This suggests the emphasis should be on long-run marginal cost pricing.

It is easy to state the basic principle of marginal cost pricing and to describe the advantages that should accrue from the establishment of appropriate prices. However, it is not so straightforward to determine in practice what marginal system costs actually are in a given situation and whether the introduction of marginal cost pricing would be practicable. We shall turn to such questions in Chapter III, below.

2. EQUITY

Equity (or 'fairness') is a highly subjective concept which has spawned its own vocabulary in the water industry. Two notions of equity may be identified: first, the general income distribution in the community, which is clearly a matter for government policy, and, second, the constitution of an equitable system of charges with regard to the services received and costs imposed on consumers. The narrower notion raises important questions about cross-subsidisation from one group or generation of consumers to another and introduces the concepts of parity and equalisation.

Various interpretations of the term 'equity' may be found in literature on the pricing of water services (United Nations, 1980; Department of the Environment, 1974). At the outset we distinguish between income distribution policy and a narrower concept of equity.

First, <u>income distribution</u>. 'Social equity' and 'income redistribution' are most frequently cited as objectives of water pricing policies in developing countries. They can give rise to 'political' or 'social' pricing of water. It is sometimes argued that water supply and disposal for householders is a basic human need, and should be provided at a low, and therefore perhaps subsidised, price to ensure that no consumers are inhibited by income considerations from enjoying the benefits (United Nations, 1980; Gilling, 1980). A contrary view which has gained ground in developed countries in recent years is that when it is decided to grant financial help to poorer members of the community it should be given through the social security and taxation systems and not by (a) subsidising generally the prices of the public sector enterprises or (b) levying on some consumers (e.g., user categories, income classes, industries or regions) charges higher than the costs they impose on the system, in order to assist others (Herrington and Webb, 1981). Nevertheless this project has found that income redistribution goals remain incorporated in some communities' water pricing systems, e.g., to foster development in agricultural and industrial sectors, to lighten charges on isolated communities and to help low-income households (examples have been reported in Australia, Belgium, Canada, Italy, Japan and Portugal: see Chapter III, below).

The question of whether the water industry and water authorities should be the instrument of income distribution policy has no 'correct' answer since the response given will depend on the chosen perception of the industry. Adoption of a <u>social service</u> model of the industry, stressing the "right to potable water" (as did the 1977 United Nations Water Conference) and the external benefits arising in the field of public health (see II.4, below), may well produce a concern for broad equity. On the other hand, a <u>business</u> or <u>public corporation</u> approach to the water services, emphasising (a) water supply and treatment as an input to the commercial production activities of industrial and agricultural enterprises and (b) the 'luxury' household uses of much public water supply and treatment evident now in the developed countries, generates a view on income redistribution which minimises or removes any responsibility of water industry and agencies. This second approach is more likely to be tempered in the case of major changes in policy (e.g., a move towards domestic metering), since even a public corporation may then need to have regard to any significant additional charges imposed on particular income groups.

Consideration of the narrower notion of equity raises a number of other issues and introduces the concepts of parity and equalisation. Parity is usually presented as a prescription for achieving the same average unit costs between different consumer groups or classes (e.g., between unmetered and metered consumers, or between the customer classes identified in the American Water Works Association Manual of Water Supply Practices: residential, commerical, industrial, fire-protection service and outside-city wholesale service (AWWA, 1983)). It is held to obtain when the revenue per unit quantity of the service supplied is the same for each group. Thus, as between domestic and industrial consumers on the public water supply, parity would exist when the income per thousand litres of water supplied to the two groups was the same, due allowance having been made for the assignment of leakage. More subtle notions and calculations of parity recognise not just one service (e.g., water supply) but a number of sub-services (e.g., base water supply, maximum-day supply, maximum-hour supply) and then apportion out the estimated average costs of each sub-service by applying unit costs of service to the service requirements of a given class. Service requirements may be expressed in various terms: e.g., for base supply the estimated annual quantity of water consumed by the class, and for maximum-day supply through the calculation of a capacity factor reflecting the ratio of the class peak consumption rate to the class average use rate (AWWA, 1983).

Equalisation is a concept open to a wider range of interpretations. If a water service is measured and an equal charge per unit of consumption is imposed on all consumers over a particular area irrespective of the costs of supply, then equalisation over that area is occurring. If, however, water supply and sewage disposal are not measured, equalisation proposals take other forms, e.g., levying on equal charge per unit of capital value on the properties concerned. It is occasionally argued that pursuit of narrow equity demands equalisation and thus, for example, that rural consumers should pay no more for their water and effluent disposal services than urban consumers with similar characteristics, even when supply costs are widely different. The most extreme version of this argument would equalise charges across the whole economy in much the same way as most letter post systems operate.

Critics of equalisation fall into two groups. First, some would insist on defining a water service with respect to not only its 'physical' characteristics (which may indeed be very similar in different areas) but also its cost of supply. Different costs then imply 'different' services. Second, critics emphasise that equalisation means a separation of the receipt of benefits from the bearing of costs and thus the promotion of strong forces pushing for the over-provision of services. Water authorities or areas that control their costs are discriminated against by equalisation while those that do not control their costs, benefit. This argument reduces to the proposition that pursuing one objective (equity) may mean the serious sacrifice of another (efficiency). Note, however, that part-equalisation schemes restricted to the capital costs associated with past commitments will cause no offence to the allocative efficiency objective (Frankham and Webb, 1977).

Four other equity concepts are sometimes used to attempt to justify the fairness of particular charging practices. First, that consumers should pay according to their ability. To be operational, this requires a definition of ability to pay, and this would often be given in terms of money income or wealth. A rather extreme example of the use of the wealth criterion might be the system of basing domestic water supply and effluent disposal payments in most of the United Kingdom and parts of Australia on rateable values, which in turn reflect property values.

Second, the benefit principle: that charges to consumers should be related to the value of the services to them. If applied trivially only to the marginal unit consumed, any uniform pricing scheme can be shown to be consistent with the benefit principle. If, however, it is meant to apply to all consumption, it is virtually impossible to apply strictly, for an individual consumer's preparedness-to-pay cannot be identified save in the individual contract negotiations that water authorities may enter into with very large industrial and agricultural consumers.

Third, it may be powerfully argued that it is fair that consumers should pay amounts related to the economic costs which their demands impose on the system and thus on the community. In other words, the results of allocative efficiency will ensure true equity.

Finally, there is the idea of 'historic equity', which refers to the fact that sometimes water services consumers will have made irreversible decisions, and therefore in a sense trapped themselves in situations on the basis of charges and costs as they were rather than as they are. It may accordingly be argued to be reasonable to discriminate in prices between different generations of consumers.

3. FINANCIAL REQUIREMENTS

Water undertakings are usually required to raise in revenue all of their operating costs and to service all or some of the debt associated with their capital expenditures ('historic' financial obligations). Recent developments in inflation accounting have led to moves towards charges providing for current cost depreciation as well as earning the opportunity cost of public sector capital ('economic' financial obligations).

As yet no reference has been made to the financial obligations, perhaps incorporated in statutes, of public and private sector water undertakings. Obviously these will have profound implications for charges. The precise form which financial requirements take is strongly influenced by differences in accounting conventions between nations and by whether water undertakings are in the public or private sectors. However, certain common elements within these requirements, which govern the sums to be recovered from charges, can often be identified. These elements include:

-- operating costs
-- historic or current cost depreciation
-- interest charges on outstanding debt
-- a financial target, and
-- taxation,

but not all elements are necessarily found within each set of national financial obligations. Examples of financial requirements imposed in Member countries are presented in section III.2, below.

Where a financial target is levied, it may be couched in various forms: as a percentage return on the net assets of the water undertaking, as a certain value for the gross trading surplus of the undertaking (perhaps to supplement historic cost depreciation) or as a specified value for the undertaking's self-financing ratio, ensuring that the trading surplus is sufficient to pay for a certain proportion of the period's gross capital investment.

In specifying a financial framework and, in particular, setting a financial target for water undertakings, Member states may have in mind a number of different objectives. For private sector undertakings a financial target may be set by a national regulatory body as a mechanism for ensuring that a monopoly is not used for exploitation, i.e., to ensure that the return on capital earned is not excessive. Where undertakings are in the public sector, the choice of financial target may reflect national policy on the role of public utilities. Such policy might involve utilities earning a low gross return on capital, perhaps sufficient to cover only nominal interest payments and the funds that should be retained in the utility to replace its assets at their current cost. Alternatively, policy may be directed towards public utilities making a higher return to cover the full opportunity cost of capital funds and therefore perhaps to help to reduce public sector borrowing requirements and taxation. An objection to using financial targets in pursuit of these latter objectives is that an inequitable burden may be placed on current customers: they may be made to subsidise, through high charges, investment required to meet the needs of future users. The strength of the objection depends on a judgement of the appropriate opportunity cost of capital for public sector undertakings and how that compares, when translated into a similar form of target, with the existing level of financial target set for water undertakings.

One of the most comprehensive surveys of alternative approaches to the revenue policy of a water authority is that presented by Pollett (Victorian Water and Sewerage Authorities Association et al., 1985). He distinguishes between 'Accounting', 'Economic' and 'Business Enterprise' approaches to revenue generation. Highlighted in the survey are the different treatments that may be accorded to depreciation and interest rate costs (broadly, historic actual or the opportunity cost of capital) and the different degrees of self-financing that either are targetted or simply emerge.

It should be noted that financial obligations are necessarily framed in terms of <u>financial</u> (or accounting) costs whereas our early discussion of marginal costs (II.1 above) was based on a wide definition of opportunity costs (I.2 above). A framework of financial obligations, however, may be used to transmit messages on opportunity costs both to the management and to the customers of water undertakings. Reconciliation of the efficiency and revenue generation objectives is considered in III.7 below.

4. PUBLIC HEALTH

Charging systems should not be designed or operated such as to put public health in any significant danger.

It is sometimes claimed that in relation to water supply in the domestic sector and to sanitation in all sectors charges must not be such as to force or induce consumers to economise unduly in the use or disposal of water to the detriment of their own and the public health. Excessive charges (the imprecision of which term is readily admitted) should therefore be avoided for

  i)   Connecting new consumers to the public water supply;
 ii)   Use of the public water supply;
iii)   Connecting new consumers to a hygienic sewage disposal system (piped or otherwise);
 iv)   Use of the adopted sewage disposal system.

An extreme version of this view is that water supply and disposal should be free at the margin. In this way the 'threat' to public health can be said to be minimised, and in some countries this has been used as an argument against domestic metering. But the corollary of this is that a water authority must raise its income from the domestic sector by some other means ... and all other means may readily be shown to be inequitable or impracticable. It has therefore been argued that a two-part metered tariff could be designed to include a block, sufficient to cover a household's entire 'basic' requirements, where the charge would not be related to intensity of usage and might even be free. In such circumstances it is claimed that the marginal charge would have no effect on public health because it would not apply to the basic block of consumption (e.g., Department of the Environment, 1974; para 4.16). Such an argument could be fallacious. In practice, any relatively small 'basic block' of consumption would be used up early in the billing period and the consumer might then reasonably perceive the marginal charge for all units consumed, 'basic' or otherwise, to be positive for the rest of the period.

We should also note the ironically opposite argument that a public water supply service providing 'free' marginal units can cause public health problems in sewerage and sewage disposal, if available capacity is unable to cope with the perhaps large and rapidly increasing volume of effluents that might result in the absence of any non-price demand management policies.

5. ENVIRONMENTAL EFFICIENCY

Environmental efficiency constitutes an elaboration of the allocative efficiency objective described earlier. When a pricing system is in operation, the rational use and preservation of the environment requires that all the social costs of providing the water service are reflected in the price. If production activities give rise to liquid effluents with unpredictable but perhaps serious environmental consequences, then direct controls are needed.

There is a need to recognise that the aim of water resource management is not simply to provide water of sufficient quality and quantity. The water resource additionally has ecological and recreational values and these need to be reflected in a pricing system.

A charging system should be such as to promote a sensible use of the environment, meaning that prices should ideally reflect the true and complete social costs of making available the service under consideration. This applies especially to resource depletion. For example, charges for water abstractions pumped from an aquifer should reflect the costs of 'maintaining the system intact' as well as operating costs. Quantitative resource depletion - the lowering of the water table - would thus be valued by the costs necessarily incurred to replace the water abstracted with a supply of equal quality, either by direct replenishment or by suitable augmentation of some other part of the supply system.

Identified and measured damage costs should be evaluated and reflected in prices where this is feasible and where it has been decided not to try to counter the damage through regulations.

Unmeasured environmental costs should also be reflected in the tariff when they are significant. To those who claim that unmeasurables cannot be reflected in water pricing schedules, the answer must be that for many years various equity objectives, also essentially unmeasurable, have been influential in tariff design.

Wholly unpredictable or potentially disastrous environmental consequences of water service production or consumption activities cannot be handled through pricing. They must be dealt with through amendments to, controls over or, in extreme cases, prohibition of the activities.

6. CONSUMER ACCEPTABILITY AND UNDERSTANDING

The charging system must be comprehensible to consumers and command broad acceptance among them.

For acceptability, a charging system must be perceived by consumers to reflect at least 'rough justice'. It must also therefore be able to be understood. A complex multi-part tariff for households involving seasonal and time-of-day differentiation triggered by overall system loading, might be a price designer's dream. To consumers, however, 'nightmare' might be a better description. Misunderstanding could mean the 'right' price signals generating the 'wrong' reaction; in reality, therefore, they would be the wrong price signals.

7. ADMINISTRATIVE COSTS

A charging system must not impose large administrative costs on a continuing basis unless there are clear gains to efficiency, equity, revenue generation or the public health.

Ideally, a charging system should impose low administration costs. These include the costs associated with:

   i) Obtaining information about the costs of provision of the service;
   ii) Obtaining data concerning the effect of price variations on demand;
   iii) The maintenance and reading of meters;
   iv) Keeping records (including any costs borne by users themselves)
   v) Billing;
   vi) The collection of charges;
   vii) The recovery of debts;
   viii) The definition and maintenance of zonal charging boundaries.

It is considerations such as these, together with consumer understanding, which limit the extent of feasible differentiation in charging of the sort referred to at the end of II.1, above. Furthermore, the costs of compliance must be reckoned with. Monitoring and policing the system will include ensuring that meters are not tampered with and that charges geared to the ownership or use of particular water-using appliances are not avoided by false reporting. Generally, the pursuit of more allocative efficiency implies that more precise monitoring of consumption is necessary and this in turn results in higher administrative costs.

8. ENERGY

In some circumstances particular regard should be paid to the energy consequences of water services charging schemes.

Following the energy crises of the last decade, attention has been concentrated upon the relationships between water production, use and reuse, and energy consumption (Young, 1976, and Speed, 1980).

An 'efficient' pricing system for energy, reflecting the long-run marginal cost of energy production and distribution, should in theory feed 'correct' price messages into consumer decisions concerning how much water to abstract or consume, how much recirculation to practice, how much pre-disposal treatment to undertake and how much effluent to dispose of. There are two routes along which price messages travel. First, a water authority's costs of water supply and sewage treatment (and thus the prices it sets) will reflect appropriately the cost of energy used in its 'production' processes. Second, water service consumers, in deciding what technologies to adopt in production (some complementary with, and some substituting for, water), will take account of energy costs in operating that technology. Optimal amounts of energy consumption should therefore result if energy and water services are priced 'correctly'.

If, however, energy is inappropriately priced or if consumers make 'irrational' responses to energy prices, then a conservation of energy objective set by government may require that special treatment be accorded energy costs in water authority pricing systems. Shadow pricing may be required, although publicity and direct regulation of energy use may also be exploited by the authority. 'Energy budgets' may have an important role to play in water- and energy-intensive activities like irrigation (Benson, Landgren and Cotner, 1979).

9. EMPLOYMENT

At times of high unemployment, governments may choose to integrate employment objectives with any charging scheme and financial target guidelines issued to water authorities.

Since Member governments have sometimes placed major emphasis on employment objectives, the links between employment and pricing should be understood. As with energy, it is the possibility of market failure - in this case, the social costs of using labour not being reflected accurately by wages paid - that causes most concern. If there is considerable unemployment in the trades and groups employed in constructing and operating water services, wages paid may overstate the opportunity costs of employing that labour. This may lead to 'wrong' price signals. It would normally be for government rather than the water authorities themselves to impose guidelines for the correction of such distortions.

10. CONCLUSIONS

The discussion concentrated so far on seven criteria to which any sensible charging system for water services must have regard and drawn attention to two resources (energy and labour) concerning which there may be particular constraints. Of the seven criteria, three will dominate the Chapters that follow: efficiency, equity and financial requirements. And of these three efficiency would normally be given the greater weight by those who believe economic instruments have a major role to play in resource conservation. Several OECD countries explicitly recognise the primacy of these three criteria in the formulation of demand management and pricing policies for water utilities. One recent consultant's report for the Australian government, in looking ahead to the year 2000, argued that,

> "...it will be necessary to devise pricing systems which promote efficiency and which are at the same time compatible with community notions of equity. At all levels of government, policy-makers will have to determine levels of public funding for water-related infrastructure...." (Department of Resources and Energy, 1983; emphasis added).

The other four criteria are not necessarily less important in all situations. Rather are they much less liable to analysis.

The final point is a note of caution. Although pricing systems should be reliable and robust, they cannot be expected alone to handle all eventualities. Thus in some situations , e.g., (i) in droughts; (ii) where metering is not practised or (iii) where toxic substances are being disposed of in effluents, non-price allocation methods involving regulations may be required to supplement, dominate or displace the price mechanism.

Chapter III

WATER PRICING IN PRACTICE - THE PUBLIC WATER SUPPLY

1. INTRODUCTION

This part of the report turns away from principles and begins to examine in detail the major water services listed at the outset (section I.1). First, consideration is given to the public water supply.

Submissions from OECD Member states suggest that in a number of non-European countries there is concern that water supplies are not being provided rationally. Thus in Canada the lack of a pricing system relating to the marginal value of water is said to have led to the overbuilding of systems, waste of public funds and sub-optimal water use practices. And in Australia the Hunter District Water Board's July 1982 proposals for a 'user-based tariff' were contrasted with the previous Hunter system (meters installed, but most consumers inside the generous 'free allowance' part of the tariff), which was described as "....no basis for economical and equitable operations.....skimping on maintenance for a decade....the rating system is collapsing under the weight of its internal contradictions."

While these countries have received their water charging legacies from Britain, three European countries appear more satisfied with their water industries. Switzerland has extensive water resources, "favourable prices", very high water consumption and, it seems, few pressures for more careful water use (indeed, domestic water use seems to have reached a ceiling of around 230 lhd in the early 1970s and to be static at that level). Denmark's large supplies of ground water are inexpensive because: (i) they are so prolific; (ii) the water is pure and thus requires little treatment; and (iii) there is no need for distribution from central treatment works (avoiding pipe costs). Similarly, water issues appear relatively manageable in the Federal Republic of Germany where "water is not a scarce good", although uses are regulated by tight legislation and price is said to be highly political. France maintains that the tariff based on marginal cost is only applied exceptionally "in its theoretically pure conception" in that country. However, certain services "implement tariff structures inspired by the principles of marginal cost pricing" in applying high prices when the use of installations is near to saturation and low prices in the contrary case.

Section III.2 describes financial requirements in practice and in III.3 a familiar and useful public utility cost classification is presented. In the two succeeding sections present pricing practices are described in detail, III.5 being devoted to the more difficult issue of capacity charges. III.6 shows how practical marginal cost pricing may be built upon the earlier cost classification and III.7 tackles the necessary reconciliation of marginal-cost-based prices and financial requirements.

2. FINANCIAL REQUIREMENTS IN PRACTICE

Most water authorities and companies still base their charging policies for the public water supply on financial rather than economic considerations. The main reason for this is that financial requirements (II.3, above) have been and remain the mainspring of rate-setting exercises. At its most thorough, this "accounting approach" (Hanke, 1975), attempting to ensure that total revenue earned balances total costs, can be a complex exercise including careful identification and valuation of all an authority's capital assets followed by the application of detailed depreciation assumptions. Annual accounting costs are thus specified, in various cost categories (III.3, below), and allocated initially to consumer groups, and then to consumers, in a manner deemed to reflect equity. The primacy of the financial requirement has meant that in most countries the water industry has been preoccupied with price/rate levels and until recently has therefore been little concerned with price/rate structures.

In the United States there appear to be two generally accepted methods used for the determination of the total revenue requirements of a water utility: the "cash basis" and the "utility basis" (American Water Works Association, 1983). In the former the emphasis is on revenue being sufficient to cover all cash needs (including debt charges and repayments), although an allowance for "normal replacements, extensions and improvements" to the system is usually included. On the utility basis, applied to all invester-owned (private) utilities and most municipal systems subject to state commissions or other regulatory bodies, a "fair return" on capital for investors is added for private institutions while the return to a public body must be sufficient to cover capital repayments and debt interest. In the latter case each regulatory body has its own rules and policies for determining total revenue requirements. In all cases it is clear that in the long-run the water utility is expected to break even.

In Denmark a 'Water Price Committee' sat over 1981-82 and in December 1982 reported to the Ministry of the Environment (National Agency of Environmental Protection, 1984). The committee provided detailed guidelines for local authorities to apply to the charging schemes of both private and municipal water supply plants. There is an underlying assumption in the committee's report that water supply plants should in financial terms break even. Non-recurring changes for consumers should ideally cover initial expenditure on the "main plant"; otherwise loans or appropriations should be used. Other expenditures to be covered are on new supply pipe and service pipe installations (fixed annual charges on consumers to be levied) and working expenses (including interest and debt repayment). The financial 'flavour' of these guidelines is clear. Similarly in France municipalities running their own supply systems try to secure an annual financial balance in their water supply and sewerage operations: "that is to say, their receipts in a year ought to cover the working expenses of the year and interest on and repayment of capital borrowed in earlier years" (French submission).

Only in two cases have examples been supplied of governments (central, state or local) stipulating rates of return on water utilities' capital assets to cover estimated opportunity cost. Such returns will, if utilities' actual (historic) borrowing interest rates are less than the required return on capital, imply increasing self-financing ratios, early repayment of debt and (eventually) increasing cash surpluses. In England and Wales a two-tier financial target has been set for individual water authorities from 1985-86: a lower rate (between 1 per cent and 2 per cent) on the current cost value of assets commissioned before 1 April 1985, but a full 5 per cent on the value of assets commissioned after that date. The first-tier percentage is expected to increase over the next few years, giving annual increases in water charges about 5 per cent above expected price inflation. In Australia, the Victorian state government is reported to be planning to aim for a 4 per cent real return on capital used by public authorities.

There is no 'correct' resolution of this issue. Whether aggregate revenue requirements are determined with respect to the economic (opportunity) costs of resources (including capital) or the overall financial costs of the utility need not significantly affect (a) prices for marginal units of consumption and therefore (b) resource allocation in the industry. This follows from the flexibility granted by standing (or minimum) charges or low-price blocks in tariff design (see below).

To the extent that a utility is in receipt of subsidies from one or other arm of government, then financial requirements are of course lowered. This question is explored in more detail in Chapter IX.

3. CLASSIFICATION OF COSTS

It has become common to classify the costs of public utilities which give rise to financial outlays into the following categories:

Customer (or Access) costs
Commodity (or Volume) costs
Capacity (or Demand) costs
Common (or Overhead) costs

This is sometimes known as the 'Hopkinson method' of cost classification.

Customer costs are those costs incurred regardless of (a) the number of units of the service actually consumed and (b) the size of the demand a consumer might potentially make on the system. They are therefore those costs associated with a customer being connected to a supply system even if no water services are consumed, and are of two kinds:

'one-off'   :   those connection and disconnection charges which cannot be recovered and transferred to other customers, e.g., the labour costs of laying a service pipe and installing a meter. Because of their nontransferability they are also known as "nonrecoverable" costs.

'continuing' : this category includes both regular impositions to do with the maintenance of the connection, reading meters, billing, collection of charges and various other consumer services; and also the costs of equipment (like meters) which can in principle be transferred to other customers should a particular customer disconnect from the system. This latter group are also known as 'recoverable' costs.

As well as varying with the number of consumers, these costs may be influenced by the size of individual connections and the method of charging adopted (taxation, maximum allowed consumption, recorded consumption, etc.).

Commodity costs are those that vary directly with the number of units consumed (e.g., litres of the public water supply) and certain other characteristics of those units (e.g., strength of sewage). They thus include pumping costs and chemicals.

Capacity costs are those incurred in the provision of resources, distribution, local storage, sewerage, treatment work, etc. Often they vary with one or other of the maximum demands made on the supply system, but this is seldom fully reflected in tariffs. It is more common to find capacity costs sub-divided into two groups: base-load-related and system-peak-related costs. Base-load-related costs are then often combined with commodity costs to form the basis of a commodity (or volume-related) charge. In principle, cost analysts could go much further, since the various peaks observed (seasonal, weekly, daily, hourly) normally have different implications for different elements in the supply system (e.g., water resources, trunk mains, service reservoirs and distribution mains respectively). Because the main consumer groups may differ significantly in their responsibility for the different peaks, attempts to specify separately the various peak-related capacity costs can be justified in the pursuit of both equity and efficiency. Examples of detailed cost allocation may be found in Hanke (1972), Feldman (1975) and Thames Water (1978).

Common costs is a catch-all term for those costs which do not vary with usage or system peaks and are not associated with a customer's connection to the supply system. They are also termed 'overhead' costs. Not being avoidable, they are probably best grouped with 'continuing' customer costs for charging purposes.

Such a classification has been used in the past in attempts to allocate all an authority's anticipated year-ahead accounting (or 'budget') costs. Note, however, that there is no reason why the capital component of these costs should be constrained to appear in 'full historical cost' terms (= historic cost depreciation + interest charges); for if a government or regulatory agency insists that a water authority charge replacement cost depreciation and specifies a rate of return on capital, then the aim would remain to allocate all the costs that must be recovered through charges.

Budget costs having been divided up into as many categories and

38

sub-categories as is reasonable, consumers would then be divided into classes, the aim being to establish parity (see II.2, above) by charging to each class the appropriate proportion of each cost category. Being fair to consumers as a class does not imply equity for each individual, however, since some averaging among individual consumers is bound to occur. What is essential, though, is that each of the classes specified should be reasonably homogeneous with respect to the nature of their demands upon the system. In this way equity may be approximately upheld.

4. PRESENT PRICING PRACTICES

Such an extended discussion of costs provides a useful introduction to a survey of present charging practices in OECD Member states.

In Australia, Canada, Norway and the United Kingdom, charges for the residential sector of the public water supply are largely based on the value of property, while in continental Europe, Japan, most of Scandanavia and the United States a combination of fixed and volumetric charges is the general rule. Table 1 shows the distribution of water rate schedules for three countries where domestic sector details are known.

In all countries the vast majority of industrial consumers have their bills based on volumetric charges, with or without an additional fixed element or minimum charge. In the commercial sector, however, charging practices vary a great deal.

The sections that follow deal with the different charging schemes found to be in operation.

Flat-rate charges

Until the end of the nineteenth century most domestic water supplied by private and public utilities in the industrialised countries was charged on a flat-rate basis. This remains the case in some countries today, e.g., in the United Kingdom (although a few consumers there have opted for domestic metering), in all of New Zealand except Auckland, in most of Australia, Canada and Norway, and in parts of The Netherlands. Flat-rate payments have been levied on various bases, including number of residents, number and type of water-using fixtures, the number of taps, number of rooms in the house, the aperture of the inflow pipe, ground area and a number of measures of property value (e.g., unimproved capital value, improved capital value, annual rental value). The practice of imposing a high minimum rate, as in Brisbane,

Table 1

PUBLIC WATER SUPPLY RATE SCHEDULES IN THREE COUNTRIES

	USA (All sectors, 1982)	Belgium (All sectors, 1983?)	Canada (Residential sector, 1983)
Fixed charge	-	5	62
Uniform volume charge	2		-
Fixed charge + volume charge	7	19	-
Minimum charge + volume charge	26		-
Fixed charge + decreasing block	4		
Minimum charge + decreasing block	56	69	34
Fixed charge + increasing block	3		
Minimum charge + increasing block	1	7	4
Fixed charge + seasonal rate	1	-	-
	100 %	100 %	100 %
No. of utilities in sample	(90)	(80)	(205)

Source: Lippiatt and Weber (1982) and Canadian and Belgian submission to OECD.

Australia (Rees, 1981), occasionally converts the rating system into a fixed charge per property for the vast majority of residents. Melbourne, Australia, already has a metering system for domestic residents with an annual reading in order to charge for possible excess water consumption. Each household's generous 'free entitlement' is proportional to the flat-rate charge paid which is, in turn, proportional to the estimated property value. Most households do not pay any excess on top of the standing charge (Ng, 1983). The Australian submission suggests that overall only 20 per cent of consumers go into excess where such schemes operate.

In principle, flat-rate systems offend against allocative efficiency, though less so if an extra water-using appliance means an extra charge. Such systems, however, are simple to administer, easily understood by consumers, provide a sure revenue for the utility and often require little policing.

Revenue is also cheap to collect. They are occasionally defended on the equity arguments that property value is a satisfactory proxy measure for (i) ability-to-pay or (ii) the water supply capacity which must be provided for a consumer (capacity costs being a major part of overall costs). The first argument may be countered by another equity proposition: that such systems frequently discriminate against the low volume consumer, especially one making insignificant demands at peak periods, while the second offends the efficiency criterion by placing all capital costs in a fixed charge. Evidence from a sample of 705 households in Malvern, United Kingdom, showed in 1968 that only 17 per cent of the variation in consumption could be associated with variation in rateable value, the basis of the domestic charge elsewhere in Britain (Frankham and Webb, 1977). A two-part flat-rate payment, made up of a standing charge and the rateable value element, would improve the correlation slightly and could be used to reflect separately the fixed consumer costs (in the standing charge). In the case of 419 households in the old Fylde Water Board, such a tariff would have increased the correlation coefficient to 0.31, i.e., 31 per cent of the variation in consumption could be associated with rateable value variation (Jenking, 1973) so all of the inefficiency and much of the inequity would have remained.

The Denver (Colorado, USA) Water district (88 000 residential consumers) served by the Denver Water Board provides details of one of the most intricate flat-rate billing systems. At least up until the late 1970s eight different factors were used to calculate the flat-rate monthly billing for a given household (in the neighbouring but much smaller Englewood district 13 factors were used). Different dollar rates were levied per room, first bath, extra bath, first W.C., extra W.C., automatic washing machine, garbage disposal unit and per 100 square feet of total lot size. However, econometric analysis showed that only 22 per cent of the variation in individual consumption for 900 metered households in 1976 in the (much larger) Denver metropolitan area could be explained by those factors which were used to determine the flat-rate bill in the Denver district (Morris and Jones, 1980). The deduction was that the complex billing schedule in the Denver district was a poor indicator of water use.

Antwerp Waterworks (AWW), supplying water to about 600 000 people, offers consumers the option between a 'fixed' tariff and a volumetric charge based on meter readings. 80 per cent of consumers opt for the fixed tariff, which includes two parts: a basic fee and a complementary fee. The basic fee depends on the type of dwelling; for a house, it costs the equivalent of $26 a year (1985) and 'includes' a first tap (cold water only). The complementary fee depends on the number and nature of the range of water-using appliances in the dwelling. The information is collected either before a new connection is made or by a periodic visit (about every seven years). If the consumer does not agree with the fixed tariff established, the only alternative is to ask for a supply via a meter. Water meters are imposed on certain customers when particular appliances give rise to consumption which cannot be easily estimated or where there is a risk of waste, e.g., large garden ($> 500m^2$), ponds, fountains, swimming pools and saunas.

An often powerful argument for a flat-rate status quo is the high cost of switching to and operating an alternative charging system. Such transition and transaction costs appear to frustrate the apparent efficiency and equity benefits of switching away from many existing flat-rate systems (but see Chapter VIII, below). Melbourne, however (see above), has seemingly managed in the recent past to combine the worst of both worlds with metering costs being incurred and mostly flat-rate payments charged.

Average cost pricing

At its simplest, average cost pricing groups together at least all the 'non-customer' costs specified in III.3, above, and divides them among the total number of units that are expected to be sold to generate a unit cost. One version of equity - parity - is here adhered to, down to the level of the individual consumer, the financial requirement is met if demand forecasts turn out correctly, but an efficient use of resources is not obtained (unless the price elasticity of demand for the public water supply is zero). The revenue of the authority, however, is at its most uncertain in that virtually all costs 'appear' in the commodity charge. An unanticipated industrial recession or an unexpectedly wet summer could thus mean serious financial losses. One of the most serious manifestation of allocative inefficiency with average cost pricing is that all consumers all the year round have to bear some of the perhaps large capacity costs incurred because of the activities of peak users.

Declining block tariff

The essence of this sort of scheme, the popularity of which has been indicated in Table 1, is that succeeding blocks of units of water are sold at lower and lower prices. Usually the tariff includes a fixed or minimum charge per billing period, related by some criterion (such as the size of the supply pipe) to customer costs and a part of capacity costs. A typical scheme is that shown in Table 2.

Three attempted "justifications" may be offered for such schedules. The first is that of the Water Rates Committee of the American Water Works Association (AWWA, 1983), which originally commended such rate structures in the 1950s. Because larger water-users, as a class (e.g., industrial units), tend to have lower peak factors with "characteristically lower extra capital requirements and related costs than do the smaller users, as a class", the Committee argue that a decreasing-block tariff would often be "in accordance with the cost of providing service to the respective classes". Such an argument is seriously deficient on a number of counts: (i) it signals to consumers via prices at the margin information about past rather than future costs, (ii) all information relates to average rather than marginal costs (so that the charges levied do not reflect the costs consumers impose on the system), and (iii) all customers who use water at peak times in reality contribute to the peak, regardless of how much they use at other times of the year and therefore regardless of their class's average peak factor.

Second, decreasing block tariffs might be designed if a utility, by charging consumers a marginal cost based price, were to be likely to fail to meet its financial requirements. For the higher-priced blocks can then be seen as a way by which the utility recovers as revenue part of what would otherwise be consumer's surplus (1). The prices of the 'earlier' and more expensive blocks should then be determined by residual financing requirements

Table 2

DENVER WATER BOARD, COLORADO, WATER RATES, 1981

	Monthly Usage (gallons)	Rate per 1 000 gallons
First	15.000	$ 0.68
Next	35.000	$ 0.58
Next	650.000	$ 0.46
Over	700.000	$ 0.42

(total costs less revenue from marginal-cost-based charges and from any capacity- and customer-related fixed charges) and not by an AWWA-type cost-allocation exercise. Strictly speaking, all consumers should be "clear" of the higher-priced blocks and be paying the appropriate marginal cost based on their marginal unit of consumption. That is a tall order, and there is no evidence that utilities have in fact introduced decreasing block tariffs with this justification in mind.

Third, an Australian submission argued that the decreasing block tariff for commercial and industrial consumers in Brisbane stems from a deliberate decision to favour business and industry in order to abstract development from other states. The report noted the waste of resources and overinvestment in new capacity that can result, as well as the inequity (in Brisbane) to small domestic consumers who all pay the same large fixed annual charge.

Declining-block rate structures are therefore not to be recommended since they encourage inefficient resource allocation and they are frequently inequitable. Different users, often in the same customer class, are confronted with different water prices. At a particular instant in the billing period, the price of water for a particular consumer will depend on his or her consumption of water at other times in the period. Paradoxical results may emerge, e.g., very large domestic users of water (usually engaging in large garden use in the summer) may end up facing a significantly lower marginal (and average) price than the low-volume user who makes very few claims on peak capacity. If the larger consumer has, as is more than likely in the example referred to, a much higher income, the resulting cross-subsidisation will be at variance with all conceivable notions of equity as well as reflecting serious distortions in the use of resources.

Increasing block tariff

Increasing or progressive block tariffs are now becoming quite common in developing countries (United Nations, 1980). The major reason for this development seems to have been the pursuit of an income redistribution objective (see II.2, above). The 'rich' (including both high-income households and many enterprises) use more water than the 'poor' (largely households), and, recognising the narrow tax base in many developing economies, the water authority may in effect be undertaking some of the work carried out by tax collecting and social security agencies in developed countries.

Our earlier remarks (section II.2) suggest that these broader considerations of equity, and thus progressive tariffs, might be of little interest to water authorities in developed countries. However, some recently-collected evidence shows this expectation is not completely borne out. Reports from Belgium, Greece, Italy, Japan, Portugal and Switzerland, and -- referring to a few instances -- from France and Canada, have indicated that progressive tariffs are in operation, and in at least the first five countries the first block of water was sometimes sold at a specially low 'social' or 'political' price (Kinnersley, 1980; Coe, 1978; and country submissions).

In Brussels, for example, 110 litres/household/day were 'allowed' to each household for 'health and hygiene', at a reduced rate. In many Belgian municipalities there are special tariffs for certain groups -- e.g., families with children, and pensioners. These take the form of either a reduction in the volumetric charge or an additional 'free tranche'.

An alternative 'explanation' of progressive tariffs is that, while on the basic quota(s) all users -- both rich and poor -- are treated alike in being 'underpriced', it is then the large users who are overcharged, producing especially impressive benefits in terms of water conservation since they are the users likely to have higher price elasticities of demand. That thesis remains to be tested.

The spread of progressive tariffs has been particularly dramatic in Japan since 1960: by 1978 52 per cent of water undertakings were using them. Table 3 shows the tariff introduced in Osaka City in November 1980 and that which had obtained in 1965. The Japanese submission reported that the progressive system has two objectives, one "to reflect the increased cost of the development of new water resources etc. on the charge imposed on consumers demanding a great amount of water" and the other "to promote the consumption reducing effect." However, the need for multiple tranches in the tariff to realise these objectives remains unclear.

Zurich is unique in Switzerland in being the only utility to have an increasing-block tariff. Up until 1968 there had been a steeply-declining block tariff together with a minimum payment (plus a fixed annual allocation). This gave few incentives to conserve water. Over the 25 years to 1970 water consumption in fact doubled. Because of increasing contamination of ground water and limits to reuse after treatment, it was necessary for Zurich to take measures to reduce radically the rate of growth.

A major innovation was the redesign of tariffs. The switch to a single volumetric charge (1968), the abandonment of the minimum charge (1971) and, finally, the introduction of a steep but simple (2-step) progressive tariff (1975) appear to have had a dramatic effect on consumption. The progressive structure introduced the idea of excess water consumption, which for a consumer with given meter size in 1975 was defined to commence at double the average 1974 consumption for consumers with that particular meter size.

Table 3

OSAKA WATER RATES, 1965 AND 1980

Date of revision	1965 m^3/month	yen	1980 m^3/month	yen
Basic charge	0-10	130*	0-10	340*
Excess Charge (per 1m^3)	11-30	17	11-20	50
			21-30	65
	31-50	22	31-40	77
			41-50	117
	51-	25	51-100	144
			101-200	182
			201-500	206
* = minimum charge			501-1000	225
			1001-	240

Source: Kinnersley (1980) and Osaka Municipal Waterworks Bureau.

The definition of 'excess water consumption' has remained the same in physical terms since 1975. For all excess consumption the price is double that of the relatively large 'basic block'. The price thus stays the same over all consumer groups (meter sizes); it is the extent of the 'low price' tranche for each group that varies. Table 4 shows what has happened to consumption.

The trends over the period since 1970 are clear: a substantial reduction in excess consumption, stability in the consumption of normal users and a 14 per cent reduction in the consumption of large users. Zurich Water Department are confident that these trends result from the tariff changes described and similar changes in the wastewater tariff in 1968 and 1971 which effectively abolished a highly regressive declining-block structure. It should be noted, however, that the industrial use of water has fallen in many countries over the period under consideration, often due to industrial recession and restructuring away from large water using industries. It is therefore not clear from the data supplied the extent to which that the tariff changes were responsible for trend stabilisation.

In Italy important changes in the national tariff system were introduced in 1974 and 1975 "to control consumption and the wasting of water" (Martini, 1980a; Martini, 1980b). In water boards responsible for 65 per cent of the country's public water supply, the following had been implemented by 1979:

-- tariff revenue was "adjusted to average consumptive costs",

45

Table 4

ZURICH PUBLIC WATER SUPPLY, 1970-84

	1950	1960	1970	1976	1984
No. of consumers in 'excess tranche'			-	2 913 (7.3 %)	1 770 (4.5 %)
Water consumption in excess (000 cu. m.)			-	2 182 (3.4 %)	704 (1.1 %)
Water consumption (millions cu. m.)					
Normal users (0-9999 m³)			32.3	32.1	31.7
Large users (10000 m³)			18.4	16.4	15.9
Total consumption	30*	40*	50.7	48.5	47.6

Source: Wasserversorgung Zürich.
          * Estimate

-- "political tariffs" (low prices) were introduced for "essential" domestic consumption (6 to 8 m³ per month per household: 200-267 litres/household/day),

-- much higher tariffs in a basic and two or three excess blocks, examples of which are shown in Table 5.

Although over the late 1970s "the development of the standard of living was held in check by the economic crisis", the reported consumption reductions are held by Martini to be in part due to tariff revision.

Increasing-block schedules have also been recommended, for the domestic sector, by the Danish Water Price Committee referred to in section III.2, "where it is particularly important to control domestic consumption" (National Agency of Environmental Protection, 1984). This approval seems to follow from the experience -- not recorded in any detail -- of a number of waterworks systems with such schedules which "probably helps to reduce the consumption of water" for domestic purposes.

Major policy shifts towards marginal cost pricing (discussed below) might further popularise progressive tariffs if significant income distribution changes would otherwise occur (as argued in II.2, above). For then water authorities might come under political pressure to maintain approximately the original 'broad equity' situation. The most obvious way of meeting such political demands without abandoning efficiency would be to price initial blocks of water at zero or low prices, but the system loses economic

Table 5

TARIFFS IN ITALIAN CITIES, 1980

		Rome	Turin	Naples	Trieste	Genoa
Reduced rate)	) all in	30	30	80	48	68
Basic rate)	) lire	97	95	130	90	135
1st excess)	) per cubic	155	115	150	114	180
2nd excess)	) metre	310	210	280	204	315
3rd excess)	)	620	305	410	294	450
Annual reduction in per capita consumption, 1974-78		1.8 %*	5.2 %	2.8 %	-	-

* 2.5 % annual reduction for domestic consumption only

Source: Martini, 1980a.

efficiency to the extent that significant numbers of consumers consume only within the low-price blocks. Again, equity has to be traded off against allocative efficiency.

It will be appreciated that those who argue it is no part of a water undertaking's task to become involved in income redistribution find cause for concern in increasing block tariffs and the potential they create for the politicisation of an undertaking's rate-fixing activities. Further criticism of increasing block systems arises from the usual inclusion of the industrial use of the public water supply. First, a firm with high consumption of water due to the nature of its product(s), and yet doing its best to economise in water use, will still face what is in effect a penalty charge. Second, two factories merging their activities for sound economic reasons could end up paying a much increased average price per litre since consumption now reaches into a higher tranche.

The same arguments of inefficiency and inequity that were made against declining-block tariffs may of course be levelled at many increasing-block schedules. Indeed, powerful criticisms may be made of all block tariffs. Strictly speaking, a single volumetric charge is required to reach allocative efficiency for all consumers and to minimise the unwanted effects of a charging scheme on the narrower notion of equity described in section II.2.

Ideally, then, applications of the increasing-block tariff would keep the first block small, allow only one other (open-ended) block, restrict themselves to a relatively homogeneous set of consumers (e.g., households), and present the idea as an exercise in general income redistribution. It is only on the grounds of the pursuit of broad equity and/or consumer acceptance that a more complex increasing-block tariff may be defended.

It is interesting that the 1970s also witnessed the introduction of 'tariff tilting', as it is sometimes known, in domestic gas and electricity tariffs. Ireland, and Massachusetts, Michigan and Los Angeles (USA) have all introduced either free allowances or 'life-line rates' for elderly consumers to provide a minimum amount of power at a relatively low price; and Japan (1974) and California, USA (1975) introduced inverted tariffs for both electricity and gas (Bradshaw and Harris, 1983).

Two-part tariffs

In many municipal undertakings, the flat-rate and average cost pricing approaches are combined to give a two-part tariff, the first part depending on some characteristic of the consumer (e.g., size of meter, number of water outlets) and the second part being a single rate volumetric charge. The fixed part of the tariff is usually presented simply as a meter rent and is often very low. Thus in Austria in the early 1970s domestic meter rents outside Vienna ranged from 5 per cent to 10 per cent of the total bill for a family of four. In other instances the 'meter rent' is much higher and represents an attempt to recover capital costs (Coe, 1978).

From the submission from Switzerland there are details of the Swiss Gas and Water Association's tariff guidelines for local water undertakings: in principle, water supply should be metered; tariffs should include a "base tax" and a volume price; and the base should amount to between 20 per cent and 60 per cent of total returns. It is argued that the volume component of revenue must be neither too low (in which case waste may occur, due to the low price) nor too high (in which case the undertaking's revenue becomes too uncertain). Guidelines are generally followed, the "base tax" usually being determined by the size of the meter and meant to cover capital services of amortization and interest with the volumetric charge "to balance the accounts". Average-cost pricing in action! Clearly it is right to include the customer costs (of III.2, above) in the fixed charge, and volume-related costs are appropriately covered by a volumetric charge. Where the other fixed costs (the capacity costs) should find a home is examined in section III.5 below.

A different emphasis is apparent in two recent developments reported in Australia. In both cases allocative efficiency seems to have been the major objective of radical tariff redesign. Prior to 1978 in Perth the system of charging for domestic, commercial and small business customers was based on (a) property values and (b) a volumetric charge for water consumed above a (very large) water allowance related to property value. The system was therefore largely based on ability-to-pay (Victorian Water and Sewerage Authorities Association et al., 1985).

In the late 1970s, however, the outlook for the Perth Metropolitan Water Board was poor; there was a need for continued heavy capital expenditure although loan funds were declining in availability and increasing in cost, per capita demands were projected to continue to increase rapidly, there was a general expectation of higher service standards and increased environmental protection, and the community was showing increased resistance to higher rates.

Thus in July 1978 a "pay-for-service/pay-for-use" system was introduced for domestic consumers. This took the form of a two-part tariff, in which a household's water charges are computed from the formula

$$T = R + p(Q-150),$$

where T is the total annual payment, R is a fixed "water rate" in dollars reflecting pay-for-service, Q is the household's annual water consumption in kilolitres and p is the price of water consumed in excess of 150 kl. The 150 kl allowance, equivalent to 411 litres/household/day, was not conceived as a 'lifeline' tariff; rather was it designed so that the vast majority of households did go into excess (now about 90 per cent) and were there confronted by an increasingly 'realistic' volumetric charge (which rose 22 per cent in real terms over 1979-80 to 1982-83). A similar scheme for the commercial and small business sectors cannot be properly implemented until the existing significant cross-subsidisation of the residential sector by these other customer classes is eliminated. Consumer acceptance of such a move will take time.

The scheme introduced in 1982-83 in the residential sector of the Hunter District in Australia, to rectify the previous flat rate structure's gross inefficiency and extensive subsidisation, was a simple two-part tariff, considered by the Hunter District Board to be "the closest practicable approach to efficient pricing". The first, or fixed, part of the tariff was intended to cover the fixed costs of the availability of the services (about 55 per cent of the Board's costs) while the volume charge was intended to produce revenue to cover the Board's operating costs (the other 45 per cent). The considerable effects of the Perth and Hunter schemes on consumption are considered below (section VIII.4).

Empirical Estimates of Price Elasticities

Present pricing practices matter in part because only under some of them will water services be expected to be allocated in an efficient manner. It is therefore profitable to examine the evidence concerning whether in fact different prices seem to make much difference to demands. Our investigations for the public water supply are divided into two:

a) The effect on demand of the metering of domestic consumers; and

b) The influence of changes in the (positive) unit price of water taken from the public water supply.

(a) will be deferred until Chapter VIII while (b) is discussed here. Hanke has summarised estimates of price elasticities for various user groups published in the United States and the United Kingdom until the end of the 1970s (Hanke, 1978). We concentrate here on extending the area surveyed to other OECD Member states and to more recent published work in the United

States and the United Kingdom, including two studies that explicitly allow for the extra 'income effects' caused by price changes under an increasing or decreasing block tariff.

Table 6 therefore summarises the studies. The result is clear-cut: with one exception (that for industrial demands in Rotterdam) price elasticities significantly different from zero have been established. But the estimates have been lower than those found before, and it now seems safer to quote price elasticities for year-round or off-peak use in the -0.005/-0.30 range rather than the -0.4 figure derived in many of the earlier studies.

Additional information was provided by country submissions to OECD. In England & Wales it was accepted that surveys on industrial demand indicated some sensitivity to charges when consumption is metered. In Switzerland, however, it was thought that price increases had only a temporary effect on water consumption (with the exception of "major consumers"). The Canadian evidence quoted "demonstrated conclusively that water demand in urban residences showed a negative elasticity with respect to price" but the suggestion in the German report is that the elasticity is insignificant. In Japan the charging system (increasing block) is claimed to have had the effect of reducing demand. The French submission tentatively associates the stabilisation of per capita domestic consumption since 1975 with increases in the price of water. Three main reasons are offered for the increases: application of "true costs"; internalisation of external effects; and expansion of sewerage and sewage treatment services; but the report stresses the need for analytical caution. Perhaps, it is suggested, really it is the price of energy acting on hot water consumption that is responsible.

Dual-supply systems

It is convenient to include at this juncture some reference to dual-supply public water supply systems, although the limited and fragmented experience of such arrangements means that virtually no empirical material is available on the effect of different tariff structures and levels.

Low-quality piped water supplies are provided by a number of statutory water undertakings in Member states. They are based on either reclaimed wastewater or poor quality river water which has been affected by upstream pollution or, being tidal, is saline. Supplies may be further divided into those restricted to non-domestic consumers and those made available to residential users for limited purposes (as well as to others interested).

In England and Wales, for example, four out of ten regional water authorities offer low-quality non-potable piped supplies in certain industrial conurbations; in recent years these have accounted for between 2 per cent and 22 per cent of overall public water supplies in individual authorities but only 4 per cent in England and Wales as a whole. In one authority measured non-potable supplies have been priced at a rate equal to 90 per cent of that of potable supplies in recent years.

Table 6

PRICE ELASTICITIES FOR URBAN PUBLIC WATER SUPPLY

Country	Location	Type of Study	Estimated Price Elasticity		Reference
Australia	971 households in 20 groups in Perth	readings over 1976-82; pooled x-section and time series	overall:	-0.11	Metropolitan Water Authority, 1985
Australia	315 households in Perth	x-section (hypothetical valuation technique)	in-house: ex-house: overall:	-0.04 -0.31 0.18	Thomas, Syme and Gosselink, 1983
Australia	metered	x-section(?)	winter:	-0.36	Gallagher and Robinson, 1977
Australia	137 households in Toowoomba Queensland	1972-3 to 1976-7 pooled cross-section and time series	short-term: long-term:	-0.26 -0.75	Gallagher et al., 1981
Canada	Urban demand eastern Canada	x-section 1960s	winter: summer:	-0.75 -1.07	Grima, 1972
Canada	Municipal demand Victoria, B.C.	time series 1954-70	winter: summer: mid-peak: year-round:	-0.58 zero -0.25 -0.40	Sewell and Roueche, 1974
England and Wales	411 firms in Severn-Trent	water-saving investment in 1972-78		-0.3	Thackray and Archibald, 1981
England and Wales	Industrial (metered) consumption England & Wales	time series 1962-80	year-round:	-0.3	Herrington, 1982
Finland	Municipal demand Helsinki	time series 1970-78	year-round:	-0.11	Laukkanen, 1981
Netherlands	Industrial demand, Rotterdam	time series 1960s and 1970s	"no price elasticity demonstrated"		Rotterdam Water Authority, 1976
Sweden	69 domestic residences in Malmo	14 readings each over 1971-78; pooled cross-section and time series	year-round:	-0.15	Hanke & de Maré, 1982
United States	2159 households in Tucson, Arizona (water use per household)	42 readings each over 42 months, July 1976- Dec. 1979; pooled cross-section and time series	year-round:	-0.256	Martin, Ingram, Laney & Griffin, 1983
United States	Domestic use in Tucson, Arizona	time series Jan. 1974-Sept. 1977	year-round (1): (log model) (linear)	-0.27 -0.45/-0.61	Billings and Agtne, 1980
United States	Residential use in 21 study areas, eastern and western United States	cross-section early 1960s	winter: -0.06 (2) summer: -0.57 (2) (east) summer: -0.43 (2) (west)		Howe, 1982

1. Price included volumetric price of sewer use and the whole tariff schedule (increasing block was assumed to change in the same proportion as 'marginal rate' changes).

2. Changes in marginal price (= marginal block rate) only, although intramarginal rate structure allowed for in demand function. These elasticities represent significant reductions on these estimated from the same data fifteen years earlier (when the intramarginal rate structure was not allowed for): -0.23, -0.86 and -0.52, respectively (see Howe and Linaweaver, 1967, and Howe, 1982).

Other industrial and non-residential urban applications (such as the watering of public parks) have been reported in:

a) Colorado Springs, USA;

b) Danish coastal areas;

c) a concentrated industrial zone in the Basse-Seine region, France;

d) sales by the Rijn-Kennemerland Water Company and in the Rotterdam Industrial Harbour area in the Netherlands;

e) Germany, for a few very large-scale works (e.g., Volkswagenwerk/Wolfsburg, Bayerwerk/Leverkusen, and the Werk Salzgitter of the Stahlwerke Peine-Salzgitter AG);

f) Japan, where pressing supply problems have induced much interest in and considerable experience of dual supplies; five industrial water works were claimed to be utilising treated municipal sewage in 1978; and

g) North West Tasmania, Australia, where examination of the cost implications of a major regional supply project produced a number of examples of industrial users and parks being able to make use of untreated supplies at considerable financial savings.

(Young, 1976; Sagoh, Aya & Funaki, 1978; Mohle, 1979)

The second type of non-potable supply, involving domestic-type usage (normally toilet-flushing or garden consumption) is rarer, presumably because of cost or suspected health dangers. Examples exist in Hong Kong, where one-sixth of total demand in 1976 was supplied by sea-water which was distributed for toilet-flushing, and in Japan, where in 1978 supplies were in operation or in prospect for about 20 000 people in residences and in 15 factories, universities and other institutions (Sagoh, Aya & Funaki, 1978). In Italy the spread of dual-supply systems has recently been recommended by the government "in order to offer consumers an alternative source of less pure water to hose gardens, or for refrigeration, etc." (Martini, a). In Rome the present non-potable network is 500km. long with an availability of $3m^3$/sec., whereas the drinking water network is 3 200 km. long with $21m^3$/sec. availability. ACEA Rome plans to increase non-potable availability to $10m^3$/sec in the next 30 years in order to reduce peak consumption of potable water (sprinkling and hosing). In New South Wales, Australia, in the 1970s six small towns introduced dual-supply systems for all domestic consumers since local river water was cheap to obtain but very expensive to treat to potable standard. Treatment cost savings outweighed the costs of an additional distribution system (Rees, 1982). The South-West Water Authority in England investigated the feasibility of a new non-potable supply of sea-water for toilet-flushing as an alternative to an expensive reservoir development but found it hopelessly uneconomic (South West Water Authority, 1975). In the Federal Republic of Germany there were dual-supply networks in several districts of a number of cities up to about 1960 (e.g., Hanover, Frankfurt and Wiesbaden) but for economic reasons they were changed to one-pipe systems. On Heligoland sea water had been used for toilet-flushing up to the second world war but it was not restored during post-war reconstruction (Mohle, 1979).

Little information is available on pricing differentials, although Snijders has reported interesting results for a number of industrial consumers in Rotterdam who switched from potable water to the city's distilled water supply (Snijders, 1980). While there is no peak pricing of drinking water in Rotterdam, the limited production capacity of the distilled water plant makes it important for demand to be as uniform as possible. This is encouraged by:

a) A special standard charge for distilled water consumption based on a maximum (peak) volume specified by the consumer -- hourly for large customers and monthly for small;

b) Increasingly severe volumetric charges if this contract volume is exceeded.

Four out of seven consumers who switched to the distilled supply were found to be using water more uniformly after the change (and two smaller ones showed no change). Overall the annual monthly peak factor has been reduced from 1.33 to 1.16. Apparently it was the penalty clause rather than the standard charge which had the greater effect. It is believed, however, that in this case the production of distilled water was and is financially unviable, in that prices are below the costs of production.

Connection charges

As public sector financial constraints have tightened in recent years the practice has become more widespread of utilities seeking capital contributions from new water service customers. These are referred to variously as buy-in charges, latecomers' fees, connection charges, tap fees or system development charges. They have been the subject of much controversy in the United States, where related proposals have also been tabled to require new customers to pay more per unit of consumption than old customers (Hanke & Wenders, 1982). In Denmark the Water Price Committee enthused over the idea of development (or 'non-recurring') charge contributions for all capital expenditures save supply and service pipes. The contributions required, however, should not be so high that potential customers are deterred from joining the system. 'Key price factors', largely determined by the relative peak day consumption of different customer groups, are then used for apportioning the initial expenditure contributions of individual consumers.

The allocative efficiency criterion is the most useful in generating an economic perspective on this issue and in this respect it is helpful to bear in mind the distinction between 'one-off' and 'continuing' costs introduced in III.3, above. When old and new customers are jointly using facilities there is no economic reason for distinguishing between them. Suppose urban growth around a city causes the water supply system to be extended, at high cost, to new sources of supply. Who is responsible for the extension: the household that has lived in the city for fifteen years or the new family that moved in this year? Because they are equally responsible for the size of the city and its water supply system, they are equally responsible for the extension of the system. The old customer's 'last' (marginal) unit of consumption is as much responsible for system extension as the new customer's 'first' unit.

What if, however, new customers impose on the supply system some costs which are specific to their supply, costs that would not be avoided even if existing customers reduced their consumption? For example, a particularly complex branch from the main distribution system might have been necessary. If this branch is not to be usefully conceived as part of the general growth of the distribution system (which would benefit further new customers in future years), then the efficiency criterion argues that new customers should be singled out for special charges to reflect the non-recoverable elements of their connection costs. These should be presented to customers as one-off, up-front charges based on the relevant costs incurred at the time of connection.

Implicitly the normative nature of much of this discussion rests on a view of equity which maintains it is fair that consumers should pay in relation to the costs which their demands impose on the supply system. An opposing concept such as equalisation ignores some or all of the variations in supply costs and would at most seek to recover an 'average' connection charge from each new customer to reflect 'average' non-recoverable costs.

While some 'mature' water supply or sewage networks have moved in recent years to the use of connection charges to such an extent that the term 'demand management' seems a more appropriate description of what is going on, in others connecting more people and enterprises to the system remains a high priority. Sometimes this is clearly in the undertaking's financial interests (e.g., if it has much unused capacity because of earlier 'wrong' investment decisions) and sometimes in the community interest (e.g., to promote public health or to restrain groundwater overpumping arising from private and often domestic abstractions). In all these cases connection costs may be waived or reduced, although clearly a community's economic or social interests may well not be served by a water undertaking with excess supplies -- or perhaps a private water company -- virtually bribing consumers to join the network. The general trend, however, is towards larger system development charges, which in effect injects private capital into the industry. In the USA the AWWA base-extra capacity methodology for determining water rates has recently been used to calculate precise system development charges for any of a variety of customer types with any of a variety of meter sizes (Barden and Stepp, 1984). As before, peak-day and peak-hour factors play a significant role in the calculations. The result is that the volumetric charge is required to cover only operating and maintenance costs, and it thus seriously understates the long-run opportunity costs of water supply provision. As capacity costs jell into customer costs so important price messages become clouded or even lost altogether.

Drawbacks of current methods of pricing

Flat-rate payments are notoriously inefficient, since the marginal price to the consumer of all units of consumption, including those contributing to system peaks, is zero. Only the high cost of introducing metering (see Chapter VIII, below) may justify the maintenance of such systems. Flat-rate payments resulting from high minimum charges in a situation of universal metering are even more difficult to justify. Other methods of payment perform rather better on the efficiency and equity criteria, although the risk of an undertaking's revenue being less than expected is greater. That risk is lessened by the adoption of a decreasing block tariff which, in the light of III.4, above, is probably the most that can be said for such a system.

All the methods described above, save for flat-rate payments, rely to some degree on the allocation of accounting costs among different categories. There are two problems here: the allocation problem and the accounting nature of the costs.

First, not all costs may be allocable; there is no 'correct' solution to the problem of genuine joint costs. The purchase price of a sheep cannot be logically divided up into wool, mutton and hide components. Similarly, "a new reservoir, for example, may at one stroke increase system capacity, permit connection to a new class of consumers, and lower unit costs of delivering the commodity." (Hirshleifer et al., 1960.)

Second, the emphasis on the allocation of accounting costs, even in a situation when the depreciation of capital goods in an inflationary environment has been properly analysed and estimated, has meant that what is reflected in the tariff are average rather than marginal costs. Certainly financial requirements are much more easily handled if the tariff is built up from average costs (per meter, per service pipe, per customer, per unit delivered, etc.). But the price to be paid is allocative inefficiency, the magnitude of which will be dependent on the price elasticity of demand.

These two problems will be confronted in section III.6. Before that, we return to the difficult question of how to reflect capacity costs in consumer tariffs. The size of this problem is indicated by the estimates from the Thames Water Authority that a rather narrowly-defined aggregate of capacity-related costs amounted to 52 per cent of total (historic) costs for water supply in 1978-79 (Thames Water, 1978) and the United States figures to the effect that as much as 70 to 85 per cent of the costs of water supply plant facilities could be ascribed to the adoption of peak-hour (35 to 50 per cent) and maximum-day (20 to 50 per cent) design horizons (Howe & Linaweaver, 1967; Hanke, 1972; Feldman, 1975).

5. CAPACITY COSTS FURTHER CONSIDERED

For this section we shall assume that somehow capacity costs in total have been satisfactorily specified. It is then possible to identify four distinct approaches to transforming those costs into actual tariff design.

Specific capacity charges

In the first approach it is stressed that the capacity of the system is largely determined by the various system peaks that the water authority chooses to plan to meet (it is assumed that decisions regarding security of supply have already been taken). Assume that the relevant peaks have been specified and capacity costs have, notwithstanding the strictures of III.4, been allocated to those peaks. Since consumers usually have, if only implicitly, a contract with the water authority to increase their demands (and hence their contribution to the system peak) without warning, it is argued that the capacity charge(s) should be determined, as far as possible, by reference to the individual consumer's actual or potential contribution to the system peak(s).

System diversity generally ensures than an individual consumer's own maximum potential peak demand and his own actual peak demand do not successfully proxy his contribution to the system peak. Consider an example. If $10 million has been established as the annuitised capacity costs of that part of a supply system associated with a peak-hour delivery of, say, 100 000 cubic metres per hour, consumers should be billed in proportion to their contribution to the system's peak-hour demand such that the total revenue (from this element alone in the tariff) aggregates to $10 million. There are then two ways forward for the tariff designer.

First, the water authority can attempt in advance to estimate or find a surrogate for each individual consumer's likely contribution to the system peak and then fix the tariff element accordingly (e.g., in the case of the peak-hour, by meter size, lawn area, whether or not certain appliances are owned (but see III.4, etc.), taking account wherever possible of historical data concerning inter-class diversity of use. In which case, (a) some inequity within consumer classes is almost certain to arise and (b) the fixed nature of the charge fails to reward any consumer who cuts down on his contribution to the peak, but (c) no financial risk (2) occurs for the authority in the sense that the 'correct' revenue total will be generated. An example is provided in the recommendations of the Danish Water Price Committee (National Agency of Environmental Protection, 1984), where it is suggested that both the regular fixed element of the tariff paid by some households and firms and the one-off initial capital contributions towards main plant and supply pipe expenditures should be based upon a "key price system", reflecting the calculated peak day to average day ratios of different consumer categories. In this way the potential contribution of a consumer to the daily peak would be reflected in the tariff, at least for 'average' consumers in each category. For example, while an apartment is an urban area would be rated with a factor 0.9, a single family house in open country would be rated at 2.0 and a farm at 2.5 (a standard "single family house in urban area, own service pipe" is 1.0).

Second, in an attempt to enhance equity and introduce some allocative efficiency, the authority could try to charge consumers according to actual use of the system at the system peak. For a peak season few problems are posed, since more frequent readings and a volumetric seasonal supplement on the commodity charge are all that is required. Note that the lack of 'fixity' in such a revenue base means that the authority's financial risk is greater and minimum charges may therefore be needed to restore the risk to an acceptable level. For maximum-day, day/night and maximum-hour peaks, however, very sophisticated metering technology is required to charge consumers according to actual use at the peak. Examples exist of night/day tariffs (see below) but the unpredictable maximum-day and maximum-hour peaks pose much greater problems. Complex meters, with advanced recording facilities, could record and therefore establish ex post facto customer contributions to system peaks, and charges could be levied accordingly. Ex post equity is thereby achieved but efficiency gains are doubtful since any adjustment of consumption is lagged. Pre-determined peak-hours and peak-days, announced in advance by the water authority, would presumably result in peak-shifting: no help to allocative efficiency and a distinct loss in equity. Thus even more complex meter technology, with instantaneous readout feedback to a consumer providing information on aggregate system demands, would be necessary for a true responsive pricing system to be in operation.

Feldman (1975) reported some years ago of research and development along these lines, but only recently has it become a commercial proposition to link the necessary components: (i) a Remote Output device, whereby a remote register can electrically access the meter, (ii) a Multirate Unit, an electronic device displaying to the consumer water consumption and rate (price) information for different tariff periods (weekends, certain hours, particular months of the year, etc.) and (iii) one-way broadcast communication facilities to consumers ("Radio Teleswitch"), permitting tariff times and prices to be easily reset. In the United Kingdom in early 1985, installation and capital costs of the necessary units (if produced on a large scale) plus cable and meter modification were estimated at between £100 and £200 per consumer with a running cost of £10-£15 per year. Benefits from consequent reduced peaks would depend on the marginal costs of the supplies and demand elasticities (information from United Kingdom Water Research Centre, private communication). The only practical example known of the measurement of maximum flow and consequent penalty charging according to the maximum recorded consumption was of instrumentation (cost then 15 000F) being introduced in the Dunkirk industrial zone in 1974 (Vermersch, 1974). No further details are known.

Appliance charges

The second approach to charging for peaks is to concentrate attention, and thus the charges themselves, on the appliances which give rise to the actual or potential peak(s), e.g., air-conditioning systems, lawn sprinklers, garden hoses and fire protection (for industrial and commercial properties). We shall omit consideration here of the last-named, since it poses wider social problems. There are three possible ways of charging for other appliances: (a) at point of sale, (b) at point of ownership and (c) by use.

If the ownership of individual appliances were found to be consistently related to the use of additional amounts of water at certain times of the year (or week, day, etc), then a charge based on ownership would allow the consumer to take account of the cost of the water (which the appliance would consume) when deciding whether or not to buy the appliance. To avoid the difficult exercise of checking on ownership within the home, a system of charges at point of sale could be considered, the revenue to accrue to the water authority in whose area the sale took place.

Monitoring of an appliance used outside the home (e.g., in the garden) is more feasible, and two possibilities arise: calculating a periodic charge to be paid by the owner and corresponding to the average use of such appliances (therefore spreading the appropriate accounting costs over all owners), or metering those consumers with such appliances (or just the appliances themselves) and levying a customer metering charge and a volumetric charge sufficient to recover anticipated costs. Point of ownership appliance charges were common in England and Wales for garden hoses and lawn sprinklers before the water industry's reorganisation in 1973, but many were dropped in attempts to rationalise changing policies under the larger regional authorities. Recently the South-West and Welsh water authorities in the

United Kingdom have decided to meter all domestic consumers with lawn sprinklers.

Because of the heterogeneity of industrial water-using equipment, appliance charges are not (with the possible exception of fire protection) easily related to industrial use of the public water supply. In any case that sector's contribution to most of the peaks that have been considered in this section is not usually very significant.

Volumetric charges

The third approach to coping with capacity costs is to turn, in the face of the many difficulties and constraints presented by the two avenues already considered, to the use of volumetric charges wherever possible. Thus if the regular diurnal peaks impose significant costs on the system, they may be partly submerged in a general night/day classification and a night/day tariff may be introduced. And the peak-day and peak-hour problems, usually occasioned by summer use, would be hidden in a seasonal volumetric tariff. In each case some peak-shaving would occur, leading to efficiency gains (which might be considerable), but inequity would remain to the extent that, for example, peak-hour and peak-day system costs are allocated to all seasonal use (and therefore all users). Both volume-related and base-load capacity costs would thus be covered by the normal year-round volumetric charge, and the summer volumetric supplement would be set to recover all the additional capacity costs. Again, minimum charges may be needed to reduce the water authority's financial risk.

The costs and any non-financial difficulties associated with the introduction and operation of seasonal and other temporal tariffs may rule them out. In such a situation, the 'best' rate schedule from the point of view of conservation is a two-part tariff incorporating in the fixed charge only 'customer' and 'common costs' (III.3, above) as well as any adjustment (positive or negative) for the authority to raise necessary overall financial requirements. The centrepiece of the tariff in this situation should be a single volumetric charge reflecting forward-looking, incremental capacity and commodity costs. That is, the volumetric charge should be based on the long-run marginal costs of supply (and perhaps disposal: see section IV.1, below), and fixed charges would be designed to cover remaining revenue requirements. If the fixed charges that result render, in the aggregate, the utility's financial risk as unacceptably high, then a system of minimum charges would have to be superimposed.

A few examples are available, however, of the application of special volumetric pricing to recover system peak costs. In Antwerp sophisticated meters with sealed electrical change-over time switches of the type more familiar to electricity consumers are installed to encourage heavy industrial consumers to shift demands from day to night. This is to spread consumption as evenly as possible over 24 hours and therefore minimise the water undertaking's storage costs. Night consumption is charged at half the day rate. Individual large industrial consumers (taking more than 18 000m^3/year) also negotiate with AWW, the Antwerp undertaking, on a maximum extraction rate, enforced by an adjustable but sealed resistor placed close to the meter. When consumers join the system or require an increase in their peak hourly flow (necessitating the resetting of the resistor) they pay a one-off capacity contribution geared to the peak flow requested, with reductions of up to 50 per cent if a specified annual consumption of up to 80 per cent of the maximum annual flow is guaranteed. The general night tariff period is 9 p.m. to 7 a.m.; this can be utilised provided that an annual minimum consumption of 18 000m^3 at day tariff is guaranteed. Indeed, the relevant period can be extended by one hour at either end if the firm guarantees a yearly minimum consumption of 80 per cent of the highest possible annual flow (which is 24 x 365 x contracted peak hourly flow). Further 'refunds' (with the effect of a decreasing-block tariff) are granted on quantities delivered at the day tariff of more than 18 000m^3/year (up to 25 per cent off) and of quantities delivered at night in excess of 25 000m3/year (up to 50 per cent off). What has been the effect of this complex tariff structure? Considerable, it appears. The ratio between peak hour and average hour industrial consumption was reported to have been reduced from 1.6 in 1963 to 1.2 in 1974 and further to 1.1 in 1985 (Vermersch, 1974; Coe, 1978; Dirickx and Tritsmans-Sprengers, 1985). Almost 60 per cent of the consumption of large industrial firms now takes place between 9 p.m. and 7 a.m.

In 1973 a metering system was introduced in the Le Havre region which measured separately volumes supplied at a flow rate below a 'maximum' contract rate (and charged at a basic rate) and at any higher flow rate (charged at an augmented rate). The cost was approximately 10 000 F. A similar system costing 13 000 F and differentiating between night and day consumption below or in excess of contract flow rates was being used in an industrial zone in Marseilles at the same time (Vermersch, 1974). No results of these two systems are known.

Seasonal charging systems

The potential benefits from introducing successful peak pricing are well illustrated by this claim from Melbourne, Australia:

"Large pipes are required to supply water on high consumption days, of which there are only 10 or 15 per year, so far most of the year the flows are relatively small and smaller pipes would be adequate. Reducing the water use by 10 per cent on these 10 or 15 days per year would defer the need to install further large pipes and would save $6 million per year in Melbourne alone."

(Langord and Heeps, 1985)

What, then, is the evidence on the existence and effects of seasonal charging? It has proved, in fact, surprisingly difficult to find evidence relating to Europe, although the United States provides an increasing number of examples.

In Germany peak demand tariffs are very rare and only applied to industrial use; in the Swiss submission, the Swiss Gas and Water Association, in their guidelines for undertakings, can envisage a summer/winter tariff distinction "if circumstances justify such a measure" (but no example is known), while in Australia the Hunter District Water Board announced their intention (in July 1982) eventually to introduce seasonal charging. But it thought the use of much more costly meters to reflect peak hour or peak day consumption would not be worthwhile. There are a few instances in France and the Netherlands of seasonal tariffs, usually in tourist/resort areas. In one example reported in France, the volumetric charge is 20 per cent higher between 1 May and 30 October each year than in the other six months. The effects on consumption are unknown.

Table 7 summarises the known evidence on effects, all from the USA. The reported consumption reductions seem to be significant, although no ex post financial or economic appraisals of the tariff innovations are to hand.

Two varieties of seasonal rate schedule are found in the US: first, simple winter/summer differentials, applied either to block tariffs or to a single volumetric charge, and, second, a summer 'excess charge' which is applied to summer consumption greater than (a) winter consumption, (b) winter consumption plus a percentage allowance, or (c) a given quantity per consumer. The first type has certain advantages from the administrative and efficiency viewpoints, especially if there is a single volume charge, since in that case all water purchased during the summer is priced at a certain rate. The second, however, may seem superficially more equitable since consumers with no summer peak of their own are not penalised. But is this fair? The fact is that all marginal demands in the summer impose extra costs on the supply system, whether they emanate from the all-year-round constant-rate consumer or the hotelier, for example, with a busy summer season and concentrated water use. We have returned to the question of what is equitable!

The problems in implementing summer rates may be considerable. Apart from the extra administration, more revenue estimation facilities, etc., lack of coincidental meter-reading produces a definition of 'summer' that will vary over consumers. But perhaps the major problem is that of revenue stability, for large increases or decreases in revenue can result from the changes in sales volumes caused by fluctuations in climatic conditions. To reduce this problem, one American system imposed high fixed charges (to raise 45 per cent of the utility's revenue requirement) but even with a summer/winter price differential of 1.5 : 1 the summer rate then turned out to be lower than the previous annual rate (to balance total revenues and costs). Better to fix the volume charges first and then ensure that financial risk is not too low through a system of minimum charges (Russell, 1984; Mann & Schlenger, 1982).

Simulations for Washington, D.C. (Davis & Hanke, 1971) and Victoria, Canada (Sewell & Roueche, 1974) have shown how seasonal pricing via differentiated commodity charges can defer system expansion. By (a) estimating price elasticities of demand for the two areas, (b) stipulating a two-period split (6 months each) for Washington and a three-period split (7 months off-peak, 3 months on-peak and 2 months mid-peak) for Victoria, and (c) calculating marginal costs in a fairly crude fashion from recorded historic accounting costs, 'seasonal' prices were estimated and 'imposed' upon recent years' data to give simulated system demands, revenues, etc. The

Table 7

SEASONAL TARIFFS IN THE UNITED STATES

Water Authority	Year of Introduction	Details of Scheme	Outline Effects	Reference
Fairfax County, Virginia	1974	Peak use charge of $2.45 per 1000 US galls (3 785l) on all 2-qtr. summer use greater than 1.3 times winter qtr. use. Ordinary commodity charge = 70c. per 1 000 US galls.	1974-80: 6 % fall in consumption; Peak day ratio fell 1.63 to 1.4 (similar climate) 1974-77	Griffith (1982)
Dallas, Texas	1977	For each consumer, surcharge of $0.05/m³ (=31 %) in all June to September consumption above 12m³ per month.	1977 peak day ratio fell from 1.97 to 1.8 (climate "more adverse") Peak day demand fell by 12 % on previous 5-year maximum.	Rice & Shaw (1978)
Tucson, Arizona	1977	High-tranche summer tariff 60 % above low-tranche summer and constant winter tariffs (residential). Differential 40 % to 70 % summer mark-ups for other consumer groups.	1980 peak day demand down 25 % on 1976. Average daily demands down by 15 % (but simultaneous 'beat the Peak' campaign in operation).	Zamora, Kneese & Erikson, 1981
Santa Fe, New Mexico	1978	Summer/winter differentials of 5 % to 70 % superimposed on existing decreasing-block rate.	1975-79: average daily per capita demands declined 16 %.	Zamora, Kneese & Erikson, 1981

Washington study indicated that the use of seasonal prices would have led to an 8.3 per cent decrease in peak demands (6 per cent in Victoria), a 2.6 per cent decrease in total demands (7 per cent increase in Victoria) and a 1.2 per cent decrease in total revenue (5.3 per cent in Victoria). Further additions to the cities' water supply systems would therefore have been postponed.

Emergency pricing

Drought surcharges to recoup revenue after a short and severe drought or during a prolonged one are not uncommon (e.g., Larkin, 1978), but a more interesting question is whether any element of rationing by price is acceptable in extreme shortage situations, most obviously affecting piped water supplies. One example of this is known and has been summarised by Kinnersley:

"When Denver had a drought in 1976, restrictions were placed on garden watering, limiting it to once in three days. But for $15 each, keen garden-waterers could buy a permit to exceed this quota. In fact about 7 000 households did so, and the extra revenue their payments produced (over the metered price which continued to operate) met the cost of the whole economy campaign which for the period in question kept total consumption just below 80 per cent of the five year rolling average. This shows legal restriction, pricing and publicity all working together. The result, I suggest, is more satisfactory than restricting the keen garden-waterers to the same level as the less keen or than building more dams so that the less keen could water more if they wanted to do so."

(Kinnersley, 1982, quoting an article in the Journal American Water Works Association, vol. 70, no.2.)

The scheme quoted seems to provide an excellent example of efficiency, equity and revenue requirements being satisfied simultaneously, although its scope would appear to be evidently limited to relatively low-value 'luxury' uses of water, typically in the domestic sector and normally outside the house.

Interruptable service and long-term contracts

Finally, mention should be made of a very different method of managing peak demands on the public water supply. Interruptable supply contracts would grant financial concessions to large consumers in return for agreement that the water authority should terminate or significantly reduce quantities supplied if it was found necessary. Such restrictions would be occasioned by 'normal' peak demands or by emergencies (droughts, breakdowns, strikes, etc), and the contract might thus have a specified time period written in within which service reduction might occur (e.g., July/August) or it might be open-ended temporally. Capacity charges would be reduced for those consumers agreeing to such contracts, as is sometimes the case with electricity supply, since the water authority would be able to hold reserve (peak) capacity at a lower level than otherwise. Contracts may therefore serve to 'reduce' demand by up to a definite amount, which constitutes useful planning information for the supplying authority, but they provide less clear information than does peak-demand pricing on how much consumers value extra capacity. At present no example is known of the use of these contracts in the water industry. No doubt there would be strong technical objections to unannounced interruptions, but the contention here is that voluntary agreements allied with an early warning system may make economic sense for both the water supply undertaking and certain industrial users.

It is also appropriate to refer here to general long-term contract pricing by an undertaking in order to reduce the risk that an unexpected turn-down in demand will lead to an overall revenue short fall. Such arrangements are characterised by an agreement on the part of the customer to buy at least a certain amount of water at a price fixed in advance (in either money or real terms) and for a certain number of years in the future. In return the undertaking may grant a preferred status to the customer, covering periods of shortage, etc. Again, no details have been submitted of the operation of such contracts, although they are known to exist in Member countries.

6. MARGINAL COST PRICING IN PRACTICE

In III.4 the serious problems of <u>allocating</u> some costs were noted. It is not necessary, however, to throw the cost <u>classification</u> of III.3 overboard. Indeed, arbitrariness is reduced (if not eliminated) if the classification spelt out in III.3 is used as a method of organising sensibly how marginal cost pricing might be put into practice. For this task it is essential to conceive of customer, capacity and commodity costs (and any further sub-divisions) as the <u>dimensions</u> of total costs and to concentrate on establishing the <u>marginal</u> cost of each dimension: the cost of connecting another customer (with, in principle, capacity and deliveries held constant), the cost of increasing deliveries by one unit, and the cost of adding another unit of capacity (or even, introducing further dimensions, the cost of an extra unit of 'maximum hour' capacity, of 'maximum day' capacity, of 'summer' capacity, etc.).

It should of course be appreciated that it will only be worth expending significant effort on the establishment of marginal costs if there is any likelihood of them being used as signalling devices to influence consumer demand behaviour. Thus the marginal cost of peak-hour capacity may be worth estimating only if, within a feasible tariff structure, it is possible to present it to consumers in the form of a price so that they make the appropriate responses in their demand for peak capacity. If that is impossible, so that a consumer cannot be confronted, either simultaneously or ahead of time, with the cost implications of any increase in, for example, his hourly demand at the time of the system peak, then the calculation effort would be unjustified, at least as far as tariff design is concerned.

How in practice is it possible to measure marginal costs in a 'systems' industry such as water (3), granted the sort of information and cost data that might be already available or made available to a water authority? The answer is due largely to Turvey (1969) and it has been developed and applied by the World Bank (1977).

In dealing with capacity costs, the essence of the methodology is to look ahead at the water authority's or undertaking's actual or potential investment programme to deal with expansion (or, much more rarely, its contraction plan) and deduce as much as possible about the extra costs that extra demands threaten to impose on the supply system (or the savings from lower demands). There are in fact a number of ways in which the extra costs may be precisely specified, measured and related to the extra demands. These are set out in Annex 2.

Although this method may be used to give price signals highly relevant to short-run allocative efficiency in the water industry (as Turvey himself has shown in a practical application (1976)), the most useful simulations have been in the establishment of a longer-run marginal cost [By the World Bank (1977) and Hanke (1981)] which takes account of, in the World Bank's approach, the next "major lump of investment" whenever it occurs, or, in Hanke's example, of the whole investment programme related to increasing water deliveries over the medium-term future (10 to 15 years ahead).

For all the Present Worth (PW) calculations, and for the equivalent annual costs calculation, a discount rate will be needed. Normally this should be the opportunity cost of public sector capital investment; it should be 'handed down' by the appropriate central or local government department (typically, the Finance Ministry) and not have to be estimated by water authority managers.

Other costs elements can be approached in the same spirit:

Customer Costs: (a) 'one-off'

One-off connection, meter installation and disconnection charges should be priced at their marginal cost (sometimes little different from the average cost) and payment should be of a capital sum of money, in keeping with the nature of the service. It can be shown, however, that capital market imperfections may inhibit system joining/leaving activities (and are therefore responsible for allocative inefficiency). A case can thus be made for annualising, or at least spreading over a number of years, connection and installation charges, effectively adding them to the fixed element of the periodic bill.

Customer Costs: (b) 'continuing'

These should be expressed as a fixed charge per customer (or per meter) per billing period and may vary by consumer group, size of connection, etc. Administrative and cost considerations will dictate the extent of averaging that is desirable across never entirely homogeneous consumer classes, i.e., the extent to which sub-classification of classes is justified. Marginal cost estimation methods are probably of less relevance here.

Commodity Charges

The methodologies of the previous paragraphs transform capacity costs into unit (but marginal) costs and may therefore form the bulk of the commodity charge, such is the capital intensity of the water industry.

Operating costs include only those costs that vary directly with water use (largely power and chemicals) and marginal running costs should thus be estimated as the anticipated increase in annual running costs divided by the anticipated increase in the annual quantity consumed (assuming system expansion is anticipated).

But what of calculation in practice? A limited number of examples are to hand. The present author has used the 'present worth difference' methodology described in Annex 2 to establish the marginal cost of wholesale water deliveries from a planned diversion of water by pumping through the Andean mountains in Peru (Herrington, 1980). Hanke used a similar method to establish the marginal capital costs of a New York water utility (Hanke, 1981). The consultants to a United Kingdom government committee set up to report on domestic water metering have also used this technique, to estimate that part of marginal capital costs due to strategic resource development by examining the costs of different strategic programmes associated with different future demand forecasts (Department of the Environment, 1985). Second, if there is no sensitivity testing of demand forecasts, then 'average' marginal capital cost (MCC) may be calculated as described in Annex 2. The United Kingdom Water Resources Board used this technique in appraising broad investment programmes and for assessing various desalination options in the early 1970s. MCC + MOC (marginal operating cost) was there known as DUC (Discounted Unit Cost) (See Water Resources Board (1974) and Water Resources Board (1972)). A similar approach has been used to establish the benefits of reductions in leakage from the public water supply (Department of the Environment/National Water Council (1980); Hanke (1980a)) and of possible decisions to install water meters in unmetered properties (see Chapter VIII, below). Third, the 'textbook' definition of MCC could be calculated by examining cost data of the next major investment expected and relating it to incremental output (again, see Annex 2). This has been undertaken in Osaka, Japan, in recent years to form the basis of increasing block tariffs (Kinnersley, 1980). In France, also, marginal cost notions have informed some tariff design. A celebrated example emphasising irrigation pricing is described in Chapter VI.

A separate seasonal unit charge (a supplement) could be estimated by similar methods on the basis of the extra capacity costs associated with extra summer use. Hanke has calculated this for the Spring Valley Water Company, New York using a present worth-type formulation (Hanke, 1981).

7. OBJECTIONS TO MARGINAL COST PRICING

It is clear that marginal cost pricing in one or other of the forms outlined above is in practice little used among the water utilities of OECD member states. It is important to understand why this is the case and, indeed, to reflect here the objections to marginal cost pricing made by some water supply utility managers and national associations (e.g., Netherlands VEWIN, American Water Works Association). For this purpose it is convenient to summarise some of the arguments contained in the 1984 report of the Committee on Pricing Systems set up by the Dutch Water Supply Association (VEWIN).

The Dutch Committee, in trying to develop "a pricing system which could in principle be used by all water supply companies and which would enable the calculation and structure of water charges to be standardised", believed that a pricing system had two functions: <u>allocative</u> and <u>cost recovery</u>. The allocative function is central to what is termed the "welfare economics" approach (=marginal cost pricing), while in an alternative approach -- "integral cost imputation" -- cost recovery in an equitable fashion is the main aim.

The committee believed that the cost imputation system was preferable since, as well as being "in principle completely fair and understandable", it also offered

> "ample scope for the allocative function....since a careful imputation of costs according to the principle of incurrence will make it possible [for the consumer] to distinguish between relatively expensive and relatively cheap patterns of consumption."

(this latter point is apparently referring to the desirability of night/day and seasonal tariffs for some larger consumers).

But the mainspring of the committee's belief in cost imputation was its opinion that a pricing system based on long-run marginal costs (LRMCs) had a number of defects:

i) LRMCs are difficult to determine

   a) the length of period under consideration (5 years? 10 years?) could affect estimation

   b) LRMC might fluctuate over time.

ii) Pricing at LRMC might mean too much or too little revenue. Thus surpluses or subsidies may occur, and these would affect consumers' incomes "in a way felt to be unfair" (to avoid such unfair effects, "prohibitively high administration costs... would be incurred in the introduction of the extensive differentiation in standing charges necessary").

iii) Charges would vary from region to region.

iv) LRMC calculations require that a water authority specify precisely its demand forecasts and desired investment programme.

v) The relevant methodology is still "new".

vi) An acceptable inflation-free discount rate has to be specified, often by an outside body.

vii) There may be problems in determining output increments and identifying marginal costs where there are joint services.

viii) The price elasticities of the demands for water are so low that marginal cost pricing gives comparatively few efficiency gains as compared with any other 'more straightforward' volumetric charge.

Each of these arguments are dealt with below:

i) It is true that there is variation in the forward-time horizon used in different calculations of LRMC. Some case studies have looked ahead 20 or 30 years (thus deriving a long-term 'average' marginal capital cost) while others have used capital programme changes restricted to the next 5 years. It would seem to be preferable to take a longer-term view, since one source of fluctuations is thereby removed and, as argued above, improved investment decisions of water consumers should result (section II.1). Not all sources are eliminated, however, since as changing perceptions of the future are reflected in sometimes rapidly changing demand forecasts -- with consequent changes in the investment programme -- it is in principle possible that variation in forward-looking LRMCs will occur. In practice the variation is probably unlikely to be very great.

ii) There are ways of coping with too much or too little revenue, and they are dealt with in III.7, below.

iii) There are in fact strong efficiency arguments why charges should vary from region to region.

iv) The requirement of explicit demand forecasts and investment programmes is salutary. Admittedly it may demand considerable effort and perhaps radical revision of traditional management practices. It may be the case that only when the (social) costs of not pricing efficiently are brought home to a utility (e.g., deterioration of system through inadequate financing resulting from emphasis on backward-looking historic costs; intervention by central government or other funding body; public rate review hearings; etc.) would there be a real incentive to consider seriously the benefits of a move towards LRMC pricing.

v) The methodology has been widely used in other utilities (especially electricity and gas).

vi) The discount rate is in any case needed for sensible investment appraisal.

vii) Joint services give rise to the same problems even when average costs are used.

viii) The question of elasticities is much the most serious of the arguments presented above. If, in fact, price elasticities are and are expected to remain quite small, then the case for marginal cost pricing is and will be much weaker. Evidence on elasticities summarised in this report suggests that for both external-to-the-house domestic and irrigation demands, the figures are of such a magnitude that allocative efficiency might be significantly and adversely affected by getting the price 'wrong'. For other sectors and types of use, however, there must be serious doubts about 'large' elasticities, and therefore a simpler 'pay-for-use' pricing objective would probably satisfy many of the concerns of the tariff designer.

8. IMPLICATIONS OF MARGINAL COST PRICING FOR FINANCIAL TARGETS

Charging policies based on marginal costs are consistent with the efficiency objective and therefore, it is sometimes argued, with the narrower notion of equity (II.2, above). What of the other main objective of pricing policy, revenue generation?

Depending on whether the water industry is characterised by increasing or decreasing unit costs as scale expands, it is easily shown that tariffs based on marginal costs may generate either more or less revenue than is necessary to cover (a) accounting requirements based on average costs or (b) an aggregate financial requirement equal to the costs of operating, maintaining and replacing the existing supply system. In this section we shall concentrate on the increasing costs (diseconomies of scale) case and therefore first address the question of who should benefit from a surplus. In the terms of section II.2 this is a concern about income distribution rather than the narrower notions of equity.

One of two courses of action is possible. The surplus can be transferred to the state or national government (perhaps in a form analogous to a commercial dividend payment) to be distributed as lower taxes, higher social security payments or subsidies to loss-making nationalised industries. Alternatively, it may be returned to (or never taken away from) the consumer of water services in the form of lower prices. Which of these two is more "broadly equitable" is a complex question depending on (a) the efficacy of subsidies on goods compared to general tax reductions/social security payments as instruments of income redistribution and (b) optimal taxation policy. The answer is beyond the scope of this paper.

Where the decision is to distribute ('leave') surpluses directly to (with) the customer, then allocative efficiency is best pursued by means of altering marginal cost prices in such a way as to minimise the resulting distortion to the allocation of resources. The necessary price reductions should be concentrated on those prices or parts of the tariff to which demand is least responsive. Thus, if two-part tariffs are in use, the reductions should normally be in the fixed part of the tariff (which might therefore become negative), since consumer behaviour - essentially, to stay in or leave the system - would be most unlikely to be significantly affected (4). If demand functions for different consumer groups have different own-price elasticities, then the price reductions should be concentrated on those groups where demands are most price-inelastic. All these solutions to the financial surplus problem will affect (narrow) equity to some degree since different consumers' bills will be reduced by differing relative magnitudes.

At the other extreme, an overriding concern for equity might well be interpreted as an equal scaling-down of both the fixed and variable parts of the tariff. The result would offend the efficiency criterion, since unit prices would transmit the wrong signals. The magnitude of the resulting resource misallocation would depend on the sizes of the price elasticities of demand of different consumer groups for the particular service under consideration and the extent of the unit price reduction necessary to remove the undesired element of the financial surplus.

Other possible solutions to the problem include reducing the fixed charge to zero and either concentrating any necessary additional revenue reduction on a decrease in the volume charge or introducing a special low (perhaps zero) price consumption tranche for 'initial' units of consumption which would otherwise have been priced on a marginal cost basis. The size of the tranche could, like the fixed charge, be determined by various criteria. If efficiency is considered more important, then the tranche could be fixed so as to attempt to minimise the number of consumers (perhaps weighted by consumption) who end up at the end of a billing period within the low price consumption range. On the other hand, a greater attention to equity would be served if the tranche was designed to be in approximate proportion to the size of the consumer's expected demand. A step function with a few large steps would ensure that this design did not simply amount to a reduction in the unit charge.

Is it possible to offer reasoned preferences as between the solutions set out in the previous three paragraphs? Generally the modification of the fixed charge in two-part tariffs is to be preferred, for the reasons given above, and in circumstances where this amendment alone preserves the fixed charge as non-negative its adoption is to be recommended.

If, however, with a zero fixed charge the surplus generated by a water authority is still too great, adoption of a negative fixed charge can give results which are perverse for equity and consumer acceptability. In particular, the water authority may end up paying some consumers to take a supply! Other solutions must therefore be sought and two cases may conveniently be distinguished. If the price elasticity of demand for the service and consumer group under consideration is at or close to zero, and likely to remain very low in the future, then only slight resource misallocation will result from the tariff change which seems to fulfil best the equity requirement: to scale down both fixed and unit price elements of the tariff equiproportionately. If, however, the price elasticity is markedly different from zero then the introduction of a zero fixed charge and a low-price consumption tranche (of a size perhaps determined by consumer characteristics) is to be preferred, with the effective marginal price for most consumers remaining as marginal cost based.

The case where tariffs based on marginal cost may generate too little revenue to satisfy the financial target is easier to handle. If the price elasticity of demand is markedly different from zero then increasing only the fixed element of the tariff is to be recommended, whereas with a very low price elasticity scaling up all elements of the tariff by the same proportion satisfies equity without serious infringement of the allocative efficiency objective.

The implicit assumption throughout this section has been that full details of the tariff are set and known ex ante. If it is decided to incorporate free allowances, however, it is possible to come closer to the objective of all consumers facing a positive price (based on marginal cost) for their marginal consumption by determining the allowances ex post. Misallocation of resources may still ensue to the extent that consumers fail to predict correctly their consumption and/or the size of their free allowance.

Chapter IV

SEWERAGE AND SEWAGE DISPOSAL

1. DOMESTIC-TYPE SEWAGE

The expense and difficulty of separately measuring waste volumes for residential users of sewers and sewage treatment is so great that sewerage and sewage disposal (S&SD) charges normally hinge upon the structure of the public water supply tariff. Usually additions to the fixed and any variable parts of the water supply tariff are made in order to cover S&SD costs, revenue generation generally being the major objective of S&SD pricing. Since for most domestic consumers the quality and composition of waste water is approximately the same, the major differences are in volumes and thus the input of water to the premises is usually a satisfactory proxy for the volume of sewage generated. Hence a S&SD 'supplement' is often regarded as a satisfactory solution to the problem (5).

The apparent price of water is thus related to the aggregate of the costs of water supply provision and the costs of the disposal of water and water-borne wastes. Usually average accounting costs determine the relationship, although recently the importance of marginal costs has become better understood (e.g., Thames Water, 1977 and IV.3, below). The theory of public goods can support an extension of S&SD facilities and services beyond that indicated by private preparedness-to-pay because of the collective benefits that accrue to the community in the form of enhanced public health and environmental quality (Milliman, 1968). An argument can therefore be made for covering some S&SD costs through collective charging schemes, which might mean taxation. The general presumption which must remain, however, is that on efficiency and equity grounds the users of S&SD services should wherever possible bear the costs of dealing with their discharges, either collectively or individually.

In Finland, for example, the Waste Water Charges Act decreed that expenditures on the public sewer system should be covered following the user-pays principle. A separate S&SD charging system was introduced in 1974; for municipal sewer systems this takes the same form as the water supply charge: a two-part tariff comprising a fixed element and a volumetric charge. For ordinary domestic users the input of water is used as a proxy for the volume of sewage generated. At the end of 1984, the volumetric charge for S&SD was, on the average, at a slightly higher level (Mk $2.70/m^3$) than the corresponding water supply charge (Mk $2.15/m^3$). The range of charges among municipalities was high: 0.30 to 6.00 Mk/m^3.

It is useful to make a distinction here between <u>sewerage</u> and <u>sewage treatment and disposal</u>. The costs incurred in providing and maintaining sewers are not influenced greatly by volumetric considerations (6) and are therefore sometimes regarded wholly as <u>capacity</u> costs in water authority cost-allocation exercises; on the other hand, treatment and disposal are regarded as giving rise to a mixture of capacity and commodity costs. This allocation of sewerage costs is given further credence if sewerage systems are "combined" in that they convey both used water (foul sewage) and stormwater. For then the size of the sewers is largely geared to surface water flows induced by peak rainfall, unrelated to the volume of foul sewage. Efficiency criteria would then indicate charging for combined sewerage according to the sizes of paved areas draining into the system, which again suggests a strong public good element and therefore some charges being levelled on municipal and highway authorities (footpaths and roads) where water undertakings operate as independent utilities. "Separate" sewerage systems for foul sewage and rainwater require, in principle, separate charging systems. In a recent development towards more rational pricing, from April 1985 all industrial and commercial consumers in the Severn-Trent region (UK) could opt to pay for surface water drainage from their properties to a public sewer through a separate area-based (square metres) charge rather than on the traditional basis of the rateable value of their property.

The effects of pricing S&SD in different ways may be hard to isolate from water supply charges, since the two are often presented together to the customer as a combined tariff. The experience of Osaka, Japan, in this respect is considered in the next section.

2. TRADE EFFLUENT CHARGES

Industrial users of public sewers present a very different picture. The chemical qualities of water-borne wastes may differ enormously from one discharger to another and in this situation a charging system reflecting differentiated charges for specialised treatment of different characteristics of wastes may well be justified (supplemented, perhaps, by regulations or 'standards': see IV.5, below).

Waste charges for industrial discharges into public sewers have become more familiar in the 1970s. In England and Wales they were introduced on a national basis in 1976, using an agreed formula (but with regional cost variations), and in the United States there is also believed to have been a significant spread of trade effluent charges beyond the 11 per cent of cities reported as imposing them in 1970 (International City Management Association, 1970).

Because the designers of trade effluent charging schemes have been motivated largely by the objectives of fairness (as between discharges) and cost-recovery (i.e., 'proper' revenue generation), the schemes have usually been designed on an accounting cost basis. At least until recently historic rather than replacement costs have been used for asset depreciation. Costs are generally split into a number of categories, for example:

$$C = R + P + B + S$$

where C = total costs;
R = reception and conveyance costs, to cover all or some part of the provision and maintenance of the sewerage system;
P = costs of primary treatment;
B = costs of biological treatment (e.g., biological oxidization);
S = costs of sludge treatment and disposal.

R is typically recovered on a volumetric basis by authorities with separate trade effluent charges, although it can be argued that it should appear in the tariff as an availability charge. P is usually presented as a volumetric charge. B is normally charged as either a surcharge per unit weight of BOD or COD (chemical oxygen demand) above the 'normal' waste strength or a charge which varies in proportion to the ratio of the oxygen demand strength of the effluent under consideration to the oxygen demand strength of all sewage in the particular area used for costing purposes. S is a charge to cover sludge treatment and disposal. B and S are often adjusted so as to be put on a volumetric basis. Hence a single volumetric charge, C, emerges.

In England and Wales such a formulation was introduced by the Regional Water Authorities in 1976, after negotiations with the C.B.I., the national employers' association. All except one authority introduced regional equalisation schemes such that the elemental costs of treatment related to the region as a whole and not to individual sewage disposal works. A study of the effect of these charges in the southern division of the Yorkshire Water Authority showed that they constituted such a small proportion of total production costs that nearly half of 28 firms surveyed reported no change in overall production costs in the transition period while the charges were being phased in (Webb & Woodfield, 1981). An increase in investment in pre-treatment and pre-discharge filtration by industrialists was noted; one-third of firms changed processes in the transition period, and the same proportion (often the same firms) planned further changes after the end of the transition period. However, firms reported that the major stimulus for increased investment in filtration equipment was the need to meet imposed consent conditions (i.e., standards). In Britain, then, the role of the charging system in regard to the efficiency objective seems unproven.

Table 8 summarises the results of four US investigations of the effects of variations in waste charges on sewage flows and sewage strengths. The studies were undertaken on various (and mostly cross-sectional) bases: intra-industry (i.e., inter-plant), inter-industry and inter-municipality. The burden of the evidence is that price increases cause significant reductions in the demand for trade effluent treatment services.

Numerous instances have been reported of industries responding vigorously to the introduction of trade effluent charges on a volumetric basis. Sharpe records reductions in wastewater flows of 70 per cent (a food processing firm), 50 per cent (an electronics firm) and 33 per cent (a steel plant) following the introduction of waste treatment charges based on "actual costs" in Springettsbury, Pa., USA (Sharpe, 1978). For Canada Tate has concluded that the experience of the ten municipalities (with 20 per cent of the Canadian population) which had introduced effluent charges by 1980 was

Table 8

ESTIMATED PRICE ELASTICITIES IN THE U.S. INDUSTRIAL
SEWAGE CHARGES INVESTIGATIONS

| | Price elasticity of | |
	Water Demand	Weight of BOD
Price of water	-0.6/-2.2	-0.4/-2.20
Price of waste strength	-0.3/-0.8	-0.05/-0.80

Notes: Price of waste strength variously measured as surcharge per 100 pounds of BOD and combined surcharge on BOD and suspended solids (BOD = biochemical oxygen demand). Price of water = price of water supply + sewage charge; variously measured as gross and net, where net price obtained by subtracting the value of the free wastes obtained per unit of added water which results from the defined normal levels in the waste-strength charges.

Sources: Ethridge, 1970; Elliot, 1972; Hanemann, 1978; McLamb, 1978.

favourable with regard to curtailing waste discharges. Thus in Winnipeg, Manitoba, "it was not until surcharges were imposed in 1958 that industry took any major steps to reduce the concentration of their wastes". They have also had "the effect of reducing the water requirements of various industries as less water is put to use in waste disposal", although no numerical estimates are presented (Tate, 1980).

In Osaka, Japan, a beneficiary-pays system for wastewater treatment had been introduced in the form of a declining block tariff as early as 1923. By 1972 the system was grossly overloaded and revenue was insufficient to pay for system maintenance; infiltration and inflow rates exceeded 60 per cent of the normal wastewater flow. The charges system was therefore put on a highly progressive basis plus an increasing block water quality surcharge whereby Biochemical Oxygen Demand (BOD) and Suspended Solid (SS) loadings were charged at an increasing rate. These innovations, together with an increasingly progressive public water supply charging scheme (see III.4, above) helped the annual consumption of water to fall by a total of 6 per cent over 1973-77. Closer examination revealed that the reduction was mostly concentrated among the large users (over 10 000m³/month). Consumption in that bracket fell by 37 per cent over the same period (Takesada, 1980). More generally, the Japanese submission is quite clear that the progressive charge system for wastewater, which had been adopted by 276 municipalities out of 378 (in 1981), has reduced water demand.

In the United States, it has been noted that the old "quantity discount" (or declining block) tariffs are being replaced by increasing-block or constant rate schedules. By 1982, a small number of US cities had begun to stipulate charges for the handling of constituents in addition to BOD, COD and

SS; examples relating to nitrogen, ammonia, phosphorus, and oil and grease were appearing in lists of trade effluent formulae.

Just one city in a 1982 66-city US survey had made an attempt to reflect peak demands in its trade effluent tariff (although only ex post facto). Oxnard, California, calculated costs for individual firms with respect to not only monthly wastewater flow, BOD discharge and SS discharge but also maximum monthly flows and discharges in the previous twelve months. Since these three peaks carried charges between two and seven times the 'current' charges, a significant incentive to reduce peaks had been incorporated.

Wastewater utilities in many countries still rely heavily on subsidies from governments and government agencies. The extent of this is explored in Chapter IX, below.

3. MARGINAL COST PRICING

An alternative basis for S&SD charges makes the efficiency objective paramount. It assumes that a firm comprehends and reacts appropriately to the pricing scheme, which relates the charges to the marginal costs of treatment. Charges therefore signal to firms the resource cost consequences of their discharge decisions. Ideally, a firm would then decide whether it would be cheaper for it (and thus the community, neglecting externalities) to pay the charges and let the treatment authority undertake the conveyancing and treatment or to make changes itself (to its production processes or by installing treatment equipment) and thereby both save money and decrease demands on the authority's facilities.

In the short-run, single-minded pursuit of immediate allocative efficiency and the presence of excess capacity in a disposal system means that only operating costs should be reflected in the charges. The usual results of such short-sightedness will emerge.

A much better case can be made out for charges to be levied so that useful information may be gleaned by the treatment authority (from the resulting demand for S&SD services) about the need for and desirability of additional investment in treatment facilities. Ideally, the authority would seek to identify the increased costs on the whole works (or system of works) were it to be redesigned to take (a) a higher volume (other characteristics of sewage held constant), (b) a higher BOD load (volume and other quality characteristics held constant), etc. Thus marginal capacity costs for 'volume-related' facilities, 'BOD-related' facilities and 'sludge-load-related' facilities in turn may be estimated, along the lines set out in III.6, above (7). Hanke and Wentworth have recently shown, with the aid of a step-by-step hypothetical example, how the marginal capital cost of a 'composite unit' of wastewater may be estimated through calculation of the increase in the present worth of system costs resulting from an increment in use. In their example, the BOD and SS loadings remain constant; hence the 'composite unit' of wastewater (Hanke & Wentworth, 1981).

Equalisation schemes for charges will normally mean that the signals themselves must be averages (of the marginal costs of extending treatment services in different parts of a region or city). It has been shown exactly how, granted the existence of a number of treatment works and the necessity for an equalisation scheme, the optimal price should be calculated in order to maximise efficiency gains overall (Webb & Woodfield, 1981).

A concern voiced in recent years has been that if different countries move at different rates (and some not at all) towards efficiency-provoking sewerage and sewage disposal charges, there will be important implications for the costs and hence the international competitiveness of certain industries. This suggests that international agreement on (a) desirable changes in pricing structures, and (b) what are appropriate financial requirements for wastewater utilities and water authorities, may be necessary.

4. FINANCIAL REQUIREMENTS

As was noted for the public water supply, average cost pricing poses problems for cost allocation (indeed, the difficult-to-allocate costs may amount to 40 per cent or 50 per cent of the total cost of a treatment works), but not for revenue generation requirements. For marginal cost pricing, however, the situation is different. Only by chance would the revenue derived from a marginal cost pricing system exactly equal the overall financial requirements. To preserve efficiency, any desired balancing of revenue and total costs should be achieved through the imposition of a lump-sum rebate or charge, which should be independent of the quantity of the discharge.

If the marginal cost of any particular dimension of sewage treatment is greater than the average cost, in principle the problem can be solved by allowing free initial blocks for dischargers (the first x cubic meters of flow, the first y kilograms of BOD, etc.) and charging a marginal cost based price for all succeeding units of volume and strength. But equity would demand that these free blocks for a major discharger should be larger than for a small discharger. However, the initial blocks must not be so large that the discharger is paying a zero marginal price. Various possible criteria may be suggested for sizing the free blocks, e.g., rateable value, size of outlet pipe, industrial classification (combined, perhaps, with the number of employees) or some moving average of past discharges. In this way allegations of inequity may be allayed and efficiency distortions minimised. Similar arguments apply in fixing the standing charges required when marginal cost pricing on all units of volume and strength fails to generate sufficient revenue to cover all costs -- the opposite case to the one just considered.

In principle, the free allowance (or standing charge) would be granted to (or levied on) each individual domestic discharger, but parity would demand that the aggregate charges for each sector (and for each 'dimension' of treatment) be proportionate to its use of that particular type of treatment. This will constrain the fixing of the allowance (or standing charge) for each household. If water supply is metered, normally all of a household's use of treatment services would be assumed to be correlated with volume, i.e., the only significant inter-household variations would be assumed to be in the volume dimension. The variable element of the charge for S&SD, added to the price of water supply, would then reflect the marginal cost of all treatment processes, transformed onto a volumetric basis with the aid of information concerning the average strength of the household sector's effluent. If water supply is unmeasured, any initial block would not appear to the individual household as a free quantity but simply as a reduction in the charge paid.

To the extent that central and local government subsidies exist (grants and low-interest loans), which is frequently the case in Member states (see Chapter IX), the financial requirements from charges are eased. The recent experience of many countries, however, is that the growth of subsidies has become unacceptable for macroeconomic reasons; to that extent the considerations of this section have become more relevant.

5. STANDARDS AND CHARGES

Economists have traditionally argued that a system of charges, such as those discussed above, is to be preferred to a system of standards, or consents, which would have to be established for each individual discharger. But the standards/charges debate is of most relevance to direct discharges into a watercourse, which are to be examined in Chapter VII. Here it is simply noted that 'consents' will be needed to control discharges to the public sewers of (a) toxic and dangerous and other highly concentrated effluents with which the treatment processes cannot cope, and (b) more ordinary effluents, if overloading threatens the sewerage system, sewage treatment plant or the personnel working there.

Chapter V

WATER RIGHTS AND DIRECT ABSTRACTIONS

In the next three parts of the report the discussion turns to what was earlier termed the 'natural provision' of water services (I.4, above). The natural capacity of bodies or flows of water to satisfy demands for water services gives rise to direct abstractions and to direct discharges of effluents by enterprises and sewage treatment works. Direct abstractions are dealt with in general terms in Chapter V following a consideration of forms of water rights ownership, while Chapter VI highlights irrigation demands and the associated allocation problems. Chapter VII covers the direct discharges of effluents.

1. INRODUCTION

Section II.1, above, sets out the theoretical requirements for the efficient allocation and development of a water service. It is now necessary to consider how different types of legal environment may facilitate the attainment of these requirements in respect of direct abstractions, and the role of charging policy in each case. Three different forms of water resource ownership will be distinguished:

-- Private water rights
-- Government or government agency ownership
-- Common (or collective) ownership

These alternative ownership forms are manifest in water laws, the development of which will have been determined by physical, economic and social factors. Laws and the associated regulations in turn affect the way in which resources are used and developed.

2. PRIVATE WATER RIGHTS

In theory a system of pure private property rights may give rise through the efficient operation of markets to an optimal allocation of water services. The requirements, however, are stringent:

77

a) certainty:	The property rights must be well-defined, i.e., certain. This will usually involve definitions of quantity, quality, location and time of use, perhaps with conditions attached.
b) transferability:	Easy (low-cost) transference of rights through purchase and sale is necessary for continuous adaption of use as demands and technologies change. In this way high-value uses would be able to bid rights away from low-value uses.
c) externality:	Spillover effects in the use of rights must be insignificant or significant externalities must be encompassed precisely in the definition of rights or a compensation system (judicial or otherwise) must be in operation to deal with spillover costs.
d) competition:	There must be competitive forces at work on both the supply and demand sides of the market.

Two contrasting traditions of private property in water are apparent in a number of western countries: the riparian doctrine and the doctrine of appropriation. How do these face up to the requirements just listed?

The riparian system links ownership of water rights to ownership of the adjacent or overlying land. On a number of the conditions above, it may be shown to be deficient once aggregate demands threaten to exceed reliable supplies.

Because the riparian doctrine does not admit priorities of use, individual abstractors have to be 'reasonable' with respect to the requirements of other riparians (thus embracing some externalities). What is 'reasonable' (and thus what is and what is not a significant externality) has to be decided by the courts but is likely to change as circumstances change (e.g., the growth of industrial or municipal demands near a river). So existing riparians can in a dynamic economy never possess certainty, even apart from climatic vagaries, as to what future quantities, qualities, etc. are represented by the rights they are holding, buying or selling. On top of this, there is the non-transferability of the water right alone. Clearly this is inefficient in an economic sense; imagine a housing market in which the household car had by law to be sold with the house. Enforced complementarity of different property rights will in general prevent property being put to the most valuable uses.

Furthermore, as economies have developed the law alone has simply not been able to cope with the rush of changing demands. Municipalities were granted exemptions from 'reasonable' use in order to organise their own water supply and effluent disposal facilities. Ironically, such exemptions in Britain often required approval from Parliament, which at about this time occasionally had to adjourn because of the smell from the deteriorating River Thames!

In contrast, under the doctrine of appropriation the water right is acquired by use. The earliest right established has preference over later users. It is possible to enhance certainty (and therefore the value of a market in rights) by the close specification of precise abstraction quantities and associated times, places and methods of diversion, as well as the precise priority standing of a user in the case of a general shortage of supplies. For example, in Colorado and other US states a special system of priorities supersedes existing priorities in times of drought. In practice, US state legislatures have chosen to impose numerous limitations on transfers (sales) of appropriate rights, usually with the intention of containing the spillover costs that might derive from a free private market. For similar reasons forfeiture for non-use of a right is often found, although free market devotees deprecate the consequent loss of the speculative function.

There are also particular problems having to do with the development of 'private property rights' water resources systems when one or more watercourses either are already, or may become, integrated into a hydrological system which is being developed by a state or regional authority, typically in the face of increasing demands. Thus a pumped storage reservoir may be proposed in order to augment the capacity of a 'private property rights' river to satisfy abstraction demands. To make a sensible investment decision the authority should take account of estimated future demands and the economic value likely to be placed upon them by abstractors. But the data available to the authority in the situation described would at best be a record of rights purchase/sale bargains struck in the past; and this would be much inferior to the information contained in authorised and/or actual abstraction time series (plus details of licence requests turned down), especially if the latter had been generated by a water authority administered pricing system reflecting the long-run marginal costs of additional abstractions. Even if such information deficiencies can be surmounted, there remains the awkward possibility following a new development of windfall financial gains for existing rights owners as certainty of abstraction is enhanced, unless an efficient and equitable financial contribution system can be devised.

The final reason why 'pure' systems of private water rights have often proved unsuitable or unpopular has to do with the expense incurred in court battles and settlements. It seems inevitable that difficulties in defining clearly 'certainty', 'transferability' and 'priority' guarantee a continuing and lucrative role for courts and judges wherever private rights are practised. Large water users tend to be wealthier than small users and thus can afford to mount more expensive legal campaigns. The possible resulting inequity is obvious.

3. PUBLIC OWNERSHIP

Direct or indirect government ownership of water resources enables the state or a suitable agency (e.g., a water authority) to:

i) Sell the rights to abstract water (perhaps for a limited period) to the highest bidders through an auction process;

ii) Distribute rights to abstractors on the basis of:

-- First-come, first-served;

-- Estimated 'social worthwhileness' of abstractions;
-- Some perception of a desirable income distributon (e.g., to 'small' farmers as well as 'large').

At the same time it may

iii) Administer a charging scheme for authorised and actual abstractions as well as or instead of levying a fee for the issue of the administered right.

In theory the water authority might then attempt to set prices in line with the opportunity costs of provision (see I.2, above) and thus assess the desirability of system expansion. However, authorities generally wish to maintain control on water use and to this end prohibit most transfers of water rights. This may detract from the efficiency of the allocation system in the short-run; a prospective abstractor, not allowed to purchase existing rights, will have to wait for an existing right to lapse or for system expansion, even though his or her (potential) abstraction may be economically more valuable than some existing ones.

4. COMMON OWNERSHIP

Two types of collective ownership are found. First, resources that have low values at the margin for abstraction or discharge purposes are often owned in common. Their value is generally not high enough to make it worthwhile to establish allocation rules or property rights. As and when resources become more valuable, perhaps because demands grow, there is a tendency for this variety of common ownership to be transformed into private or government ownership.

Second, in a number of countries cooperative ownership agreements are to be found, even when the rights are valuable. Irrigation provides a number of instances, and one of the oldest examples is the huge network of irrigation channels constructed in the regions of Alicante, Cordoba, Granada, Marcia and Valencia by the Arabs after their second invasion of Spain in the Seventh Century AD. Each 'main' water channel, together with its reticulation system, is administered by a Sindico (or committee) of farmers. The farmers are collectively responsible for the operation and maintenance of the system. Allocation practices vary among collectives, but readers will not be surprised to learn that pricing practices tend to uphold equity rather than allocative efficiency.

5. DIRECT ABSTRACTIONS: PRESENT PRICING PRACTICES

Much variation in the control of direct abstractions is apparent among Member states. In most countries licences are required for both surface and groundwater abstractions, and this obviously facilitates some regulation of demands. Exceptions are believed to include groundwater abstractions in at least one major river basin in Japan (the Yodo) and in some of the Canadian provinces and Australian states, presumably because of either strong riparian (overlying) traditions and/or high excess supplies.

Charges over and above often nominal licence fees are less common than permits. Out of ten countries for which some evidence is available (Australia, France, Japan, England and Wales, USA, Netherlands, Canada, Denmark, Greece and Switzerland) in only the first six is there known to be any system of charging by quantity of abstraction. And only in France and England and Wales is there a comprehensive charging system in each region.

Leaving aside irrigation (considered in the next part of this report), charges for direct abstractions for which information is available are of two kinds: annual charges and unit quantity charges.

A. Japan

Annual charges relate to <u>authorised</u> abstractions. In the Yodo basin, which supplies water to about 15 per cent of Japan's population, these are derived from a tariff matrix which sets out various charges on a per litre per second basis. An annual authorised abstraction which averaged out at, say, 400 litres/second would thus produce an annual charge of 400 times the appropriate matrix element. According to the OECD report on industrialised river basins (OECD, 1980) charges in the Yodo basin in the late 1970s varied according to:

a) Type of activity (e.g., industrial and mining, supplies of water to towns, other uses);

b) Location of abstractor on the river (four zones);

c) In one of the zones, the size of abstraction (greater or less than 100 litres/second).

Increasing charges were evident when comparing upstream to downstream abstractors but these differences reflected various factors in each Prefecture and apparently had no direct relationship with water quality and scarcity. Power stations faced a special two-part charge depending upon both rated output and the difference between the theoretical maximum and rated outputs, with an allowance if use (for cooling) was made of pumped storage. Charge variations by activity type and size of abstraction had no obvious economic rationale. Supplies for potable water for towns were "practically exempt from the charge" despite the higher consumptive use (i.e., lower 'return ratio'). Other users were clearly subsidising the public water supply system. There was no sign of any seasonal variation in charges.

According to its submission, Japan levies no charges for underground water, the ownership of such water is bound up with land ownership, and pumping may only be restricted by authority where subsidence, salinity or a reduction in the water table is occurring. Much more control is exercised over river water (responsible for about 70 per cent of annual water taken); there the objectives of the charging system referred to in the survey are: (a) "equity of burden"; and (b) to raise funds to pay for the maintenance of resources.

B. England and Wales

Since 1969 River Authorities and their successors, the more powerful Regional Water Authorities, have levied annual charges on all authorised abstractions, in this case on the authorised megalitre abstraction over the year (or season: see below). Water Authorities vary greatly in the sophistication of their charging systems: for example, in 1984/5, the Wessex Water Authority had 9 rates of charge (i.e., 9 elements in a tariff metrix) and the Yorkshire Water Authority 45 rates. The latter displayed the following variations (with relative charge factors in [square brackets]):

a) by use of abstraction, divided up into:

complete loss	(incl. evaporative cooling)	[150]
high loss	(spray irrigation)	[135]
general	(industrial use, public water supply)	[50]
low loss	(circulated cooling, sand and gravel washing, fish farming)	[4.5]
zero loss	(return unchanged in temperature, quality and quantity)	[1]

b) by source, divided up into:

inland water:	1st class	[10]
inland water:	2nd class)	
underground water	)	[7]
inland water:	tidal	[3]

c) by season, divided into

summer use (April-September)	[3]
all-year-round use	[2]
winter use (October-March)	[1]

In all authorities direct abstractions for domestic purposes and non-spray-irrigation use in agriculture, although having to be licensed, are exempt from charges.

C. France

Unit quantity charges relate to actual abstractions, and it is believed that the French Agences de Bassin provide some of the more detailed examples of this type of charging system. In the Seine-Normandie Agence charges were differentiated in 1976 by:

a) Zone (14 in all, reflecting quality variation)

b) Surface or groundwater

c) Season (summer: winter = 5 : 7 months)

d) Quantity of water abstracted (charges in all 14 zones) or quantity of water consumed (charges in only 4 zones)

In all 74 rates of charge were specified (OECD, 1976) but in 19 situations supplies were considered so plentiful as to warrant a zero charge. The distinction between <u>abstraction</u> and <u>consumption</u> noted at (d) performs a similar role to the 'use of abstraction' category in the Wessex scheme. 'Consumption' was usually estimated by multiplying the quantities abstracted by 'restitution coefficients', e.g.,　　　　0.93 for industrial uses,

0.80 for public supplies of
potable water
0.60 - 0.70 for run-off irrigation
and　　0.30 for spray irrigation.

For underground waters the consumption charge was 2 to 3.5 times higher than that for abstraction; for surface waters the two zonal consumption charges were at the same level as the highest abstraction charges. The 14 zonal charges were largely determined by supply and demand factors: the "main factor in calculating the rates is the local potential gap between minimum supply at low water and peak demand from agriculture, households and industry. In general the rates are high where population and industrialisation are dense and there is a risk that supply will fall short of demand". (OECD, 1980, p. 79)

The imposition of charges by the Picardy Agence de Bassin in 1970, when the water table had been found to be falling significantly at Lille, led, no doubt together with other factors, to industrial water consumption falling by 50 per cent ten years later, a possible example of the pricing of one water service affecting the demand for another. The French submission also makes clear that parts of French water law remain "archaic" with 'non-state' waters linked into essentially nineteenth century property law; this obviously inhibits rational water use. Thus for non-state waters, no water authority is able to fix maximum values of abstraction. The underground water situation is also unsatisfactory: knowledge is often poor, and there is less organisation than in the case of river water.

D. United States

A much simpler system was in operation in the Delaware River Basin in the USA, where the abstraction/consumption distinction again operated but with a much larger differential; here water consumed was charged at 100 times water simply abstracted and returned. In the lower reaches of Delaware Bay, where there was salinisation, no charges were levied, while in the intermediate zone between fresh and salt water proportional charges were levied.

E. Information for other countries

In Canada, licences are issued both to regulate abstractions and also to act as a quasi-pricing system. The licence fees appear to be "one-off", but there is no indication that supply and demand factors help to determine charge levels. In British Columbia, only surface water abstractions require licences. There is no recurring annual fee for abstractions. Similarly, abstractions are not charged for in Germany, although it is believed that a charge would have an influence on industrial water consumption. However, "difficult administrative and constitutional questions stand in the way".

The Greek report recorded problems connected with industrial use in irrigated areas, where firms are often authorised to take water from the network. Charges are complex functions of the use of the network but do not relate to the quantity of water consumed. This produces much friction between local industrialists and the supplying organisations, and disputes frequently have to be settled in the courts.

6. RATIONALE FOR CHARGES

Much evidence exists that the charging schemes just described were and still are designed in order to generate revenue to pay for water control and supply works which have made abstractions possible.

Thus in the Delaware basin the revenue from charges was used to repay expenditure incurred in building federal multi-purpose storage dams, being levied only on sections of the river affected by these dams. In Parramatta River basin, Australia (not quoted above) a charge is made specifically for the expense of maintaining the control works. The Seine-Normandie Agence charges helped to finance an aid programme which included the protection of groundwater, major interbasin transfers and the construction of impounding barrages. Note, however, that the Agences de Bassin are not the only sources of finance (most money comes from government subsidies) and that owing to interconnections between programmes for water resource development and for pollution control separate accounts for the water abstraction side are unavoidable. Abstraction charges are therefore not earmarked by the Agences; the 'balance' is that the abstraction charges plus the pollution charges (see Chapter VII, below) plus other contributions (government, local authorities, etc.) must be sufficient to meet expenditure on the Agence's 'action programme' of works.

On the other hand abstraction charges in England and Wales are earmarked; they are to meet the costs of works intended to maintain and augment the supplies available to users (such as regulating reservoirs). The accounts of the Water Authorities into which charges are paid are described as "water resources" accounts, and in 1975 an inter-authority working party recommended that the costs of reservoirs which supplied water directly and exclusively to public water supply networks should be transferred out of the "water supply account" into the "water resources account". The justification for this radical change, which on average trebled the abstraction charges in the Severn-Trent Authority, was based on opportunity cost theory. An abstractor may be on a river whose flow can be maintained at a high level to meet his needs because it receives sewage effluents, which, conceivably only exist in dry weather because the water comes out of a direct supply reservoir somewhere. Similarly, an abstractor may be able to use groundwater only because other users are using a reservoir. It therefore makes economic sense to group all storage and control works together as 'water resources'.

In raising revenue for the relevant water authority some useful incentives for making efficient use of water have of course been generated. But the mainspring of the schemes quoted seems to have been one of revenue-generation and therefore has been in the main backward-looking (to pay off past loans; but see the comments about France and England & Wales in the next section). The alternative approach -- the design of tariffs to promote allocative efficiency -- is considered in the next section.

It should also be noted that an important difference between the authorised abstraction basis of some charging schemes and the actual abstraction basis of others is the almost zero 'financial risk' incurred by authorities administering the former schemes ('financial risk' is defined in III.5, above).

7. CHARGES FOR ALLOCATIVE EFFICIENCY

It was already shown how charges should be set if optimal water allocation is desired (sections II.1 and III.6, above). In a water resource system facing increasing demands over time, the price signal ought to convey to actual and potential users the full costs of the provision of one more unit of direct abstractions. That is, long-run marginal cost pricing is desirable.

Our survey of actual pricing practices in the previous section raises two complications for this apparently straightforward principle. First, there is the question of whether it is authorised or actual abstractions (or both) that should be the subject of the charge. Assuming for the moment a regular demand throughout the year, an extra unit of authorised abstractions should be priced at the marginal capital cost of supporting or augmenting direct abstractions, since it is only extra capital costs that may become committed through an expansion of total authorisations (8). The abstractor considering a request for an increase in his authorisation will take account of both the authorisation charge and any actual use levy in deciding whether it is worthwhile to try to reserve extra units of water use; and since in total these costs should ideally be set at the long-run marginal costs of provision (including all appropriate resource depeletion and damage costs) it follows that actual abstractions should be priced at the marginal operating cost of the system (including any external costs, e.g., damage).

In theory, then, a two-part tariff would be ideal, and it would be expected that the authorised quantity unit charge would dominate the actual quantity unit charge, since the operating costs of the sort of surface and groundwater support systems referred to in section V.6 are generally minimal. Indeed, recognition of the administrative costs of meter reading and extra billing suggests that net savings might accrue were the water authority to waive the charge for actual use. Nevertheless, the main attraction (to the authority) of a charging system that relies heavily on fees related to authorised abstractions is more likely to rest with the significantly lower financial risk than with the seductive powers of economic theory.

The second complication arises because of the extreme seasonal variations in the supplies from many surface waters. It is often the case that 'winter supply' and 'summer supply' are virtually different services. Winter supplies of surface water often approximate to the 'free good' condition spelt out in I.2, above, with capital works only being necessary to support summer flows. The implications of this situation for pricing follow easily: the marginal capital costs of the support works should be presented, via prices, to the authorised abstractions of the summer users. In practice, there might be a certain low probability that the capital works would be needed in a given winter season and so some apportionment of capital costs to the winter users might be in order. For groundwater the situation is normally quite different. Because of long and often unknown percolation patterns and times the abstraction service may be best considered as non-seasonal. Often, then, no summer/winter distinction in charging for groundwater will be justified.

The charging systems described in section V.6 accord with these efficiency considerations in varying degrees with respect to tariff structures. The Yorkshire and Yodo schemes charge only for authorised abstractions, with apparently no fees for actual use, while both the Seine-Normandie and Yorkshire tariff matrices show significant premia on summer prices (but not for groundwater in France). Only the French and England & Wales authorities appear to charge for groundwater. On tariff levels, there are a few signs of marginal-cost-based pricing. Perhaps the French Agences de Bassin, with their 4 or 5 year capital works programmes, come nearest to reflecting the future capital costs of works but in those cases, as has been noted, the integration of abstraction and pollution control programmes and budgets means that abstraction fees are not set simply to reflect future abstraction costs. In England & Wales, it can be argued that the present system of financial targets plus the use of depreciation provisions based on current cost accounting (III.2, above) produces an approximation to long-run marginal cost pricing, so long as constant costs obtain. But does the precise tariff level matter, so long as the tariff structure is approximately 'correct'? The answer would be 'no' if the direct abstractions price elasticities of demand are insignificantly different from zero for non-zero prices. So it is to that question that we must now turn.

In theory we would expect the elasticities to take on a significant negative value. For surely just as there are low prices of directly abstracted water at which an industrialist does not find it worthwhile to instal recirculation equipment, there are higher and higher prices at which it becomes worthwhile to install increasing amounts of such equipment, resulting in higher and higher water re-use coefficients. Ideally, then, the price at the margin should encourage re-use when it is in the social interest, and this is best achieved by getting firms themselves to compare the marginal opportunity cost of their abstractions with the marginal cost of recycling. In practice, however, the evidence available on the effects of charges is rather limited. One year after the introduction of direct abstraction charges in England & Wales in April 1969, one researcher found evidence only of the revocation of unnecessary licences (O'Neill, 1971). And in 1980 it was tentatively suggested that in the Seine and Yodo river basins "the incentive effect [of abstraction charges] may be considerable ....They may induce thermal power stations and industries which are big users of water to avoid certain zones where abstraction rates are high." (OECD, 1980, p.80).

In the Netherlands, it should be noted, industrial ground water demands seem to have been affected not by charges but by compulsory registration of abstractions (and issuing of licenses), which occurred in the individual provinces over the period 1966-77. Investigations have shown that compulsory licensing has contributed to a reduction in industrial ground water demands (of about a quarter over 1974-78) even while industrial activity was increasing. This was largely due to a decrease in the amount abstracted for cooling purposes.

8. CHARGES AND EQUITY

It is natural to question the equity of the charging systems already discussed. Careful perusal of two of the schemes discussed above and a similar one from the United Kingdom suggests that it is only in the Yodo basin that the tariff is consciously used to redistribute income. Thus in one section of the Yodo river large authorised abstractors pay a charge twice that of small abstractors, for no obvious reason, and in the same basin supplies of drinking water to towns are "practically exempt from the charge." This is reflected in Table 9, which apportions the total revenue raised from abstraction charges in three river basins in the mid-1970s.

On economic grounds alone there appear to be no reasons for the skewed incidence of charges in Japan. Social/political factors are evidently at work and they implicitly presume a part-social service perception of the water industry (see section II.2, above).

Other equity issues arise through the extent of equalisation over a geographical area, the degree of 'seasonalisation' built into the tariff system and the treatment of agricultural demands. The degree of zoning and quality differentiation in the schemes summarised is impressive; certainly critics cannot claim here that the pricing signals are hopelessly fogged in an averaging process. There must always come a point where an even more intricate zoning process would constitute a further complication, the costs of which would not justify the extra benefits derived. Similarly for the possibility of more than two seasons; it is known, for example, that the Severn-Trent Authority examined the case for a four-season pricing structure in 1975 but discarded it as an inessential extra complication (high costs and low benefits). The related question of agricultural demands for irrigation will be deferred to the next chapter.

Table 9

REVENUE FROM ABSTRACTION CHARGES IN THREE RIVER BASINS

	Electric Power	Industry & Mining	Urban Water Supply & Others
Yodo, Japan 1975	89 %	10 %	1 %
Seine-Normandie, France 1974	3 %	26 %	71 %
Severn-Trent, UK* 1976/77	6 %	7 %	87 %

*   Forecast rather than out-turn

Source: OECD, 1980

9. NEW AND OLD CUSTOMERS

Finally, it is important to reiterate the conclusion of section III.4, that in general no case exists for distinguishing between old and new customers for a water service unless special non-recoverable expenditures are necessary to bring the new customers into the system. Thus the practice in Japan and occasionally in France of sometimes requiring new (and especially large) abstractors to finance some of the additional water control works (OECD, 1980, p. 82) has no economic rationale if both old and new customers are to make joint use of the extra facilities.

Chapter VI

IRRIGATION DEMANDS FOR AGRICULTURE

1. INTRODUCTION

Among Member states the importance of irrigation varies enormously. Comparable figures are to hand for a range of countries in the early 1980s. The proportion of all water withdrawals estimated to be routed to the agricultural sector is shown in the first column of Table 10. Generally, these figures represent slight reductions on data reported in the late 1960s. A further indication of the importance of irrigation is the proportion of irrigated area in a country's total area; both this and the proportion in the cultivated area are also recorded in the table. These series show significant increases on the data relating to the late 1960s.

The economic desirability of irrigation depends upon:

  i) Expected crop prices

 ii) Climatic and soil characteristics

iii) The price of water

 iv) The price of complementary and substitute factors of production (especially labour and energy)

  v) Available irrigation technologies and their costs.

Choice of irrigation technology and the attention paid to water losses (in transit, distribution and application) will depend upon (i) to (v). Low water prices means the value of water at the margin is low and there is consequently little incentive to improve efficiency of water use. Higher water prices may have the following effects:

a) Less water could be applied to a given crop;

b) Farmers might utilise a more efficient irrigation technology (e.g., drip irrigation) and water application practice;

c) Farmers might choose a different cropping pattern (and consequently different water use).

Table 10

IRRIGATION IN OECD MEMBER STATES, EARLY 1980s

	% of all water withdrawals to irrigation	Proportion of irrigated area to: cultivated area (%) (a)	total area (%)
Canada	7.85	1.33	0.07
USA	39.46	10.80	2.26
Japan	..	66.89	8.71
Australia	57.11	3.65	0.22
New Zealand	..	36.17	0.63
Austria	1.79	0.24	0.05
Belgium (b)	..	0.12	0.03
Denmark	..	14.71	9.20
Finland	1.35	2.55	0.20
France	19.26	6.02	2.05
Germany	0.40	4.23	1.29
Greece (c)	82.72	25.43	7.65
Iceland	..	..	..
Ireland	..	..	..
Italy	57.30	23.60	9.97
Luxembourg (b)	..	..	..
Netherlands	..	31.90	8.10
Norway	2.00	9.51	0.26
Portugal	46.86	17.75	6.87
Spain	..	15.24	6.25
Sweden	1.60	1.76	0.13
Switzerland	..	6.08	0.63
Turkey	77.85	7.70	2.72
United Kingdom	0.27	2.18	0.63
Yugoslavia	7.93	2.04	0.63

a. Cultivated area refers to arable and permanent crop land.
b. Data for Belgium include Luxembourg.
c. Data for Greece refer to total agricultural withdrawal.

Source: OECD, Environmental Data Compendium 1985

(a), (b) and (c) would free some water for use by existing irrigators on other areas of their land or for irrigation of new land; or new producers would be able to start up irrigated agricultural production.

2. EVIDENCE ON PRICE ELASTICITIES

These possible responses suggest price elasticities of demand may be higher for irrigation than for other demands. Available evidence on price elasticities comes almost wholly from the United States, and Table 11 summarises estimates for the Western USA discussed in a recent book (Anderson, 1983).

It is clear from this evidence that the demand for irrigation water in California is highly responsive to price changes. The presumption must be that price elasticities similarly significant would be found elsewhere if the research effort could be mounted.

3. IRRIGATION SYSTEMS AND PRICING POSSIBILITIES

The method by which irrigation water is delivered affects feasible pricing structures. We therefore discuss pricing under three alternative delivery systems, relying heavily for technical information on Chapter 4 of (United Nations, 1980).

A. Continuous flow

Under the continuous flow system, water flows through a canal on certain days and each farmer is free to take the quantity he wishes. Usually it will not make economic sense to measure the quantity of water used. There may be abstraction support works and there will certainly be expenses in maintaining the delivery system. Often charges will be levied on a per hectare basis and it would sometimes be worthwhile to vary those depending on the distance of the delivery point from the storage point or support works.

B. Rotation system

Under this system, water is delivered to users in turn, usually according to a predetermined schedule. Such systems are generally based on proportional divisions of an (unpredictable) annual flow rather than known volumes. Shares are sometimes tradeable, which produces a once-and-for-all pricing system based on the capital value of shares; annual operating costs and the annuitised capital costs of new and replacement works are most conveniently recovered through an annual charge based on the number of shares (or the proportion of water received). There is then an incentive for an individual farmer to irrigate efficiently, since this would enable him both to dispose of shares and to reduce his annual fees.

Table 11

CROSS-SECTIONAL PRICE ELASTICITY ESTIMATES FOR IRRIGATION DEMANDS

Author	'Average' elasticity	'Low-price' elasticity	'High price' elasticity	Area studied
I. CALIFORNIA, USA, 1960s and 1970s				
Moore	-0.65	-0.14	-1.58	San Joaquin, Calif. (linear regression)
Moore/Hedges	-0.65	-0.19	-0.70	same (quadratic regression)
Bain/Caves/ Margolis	-0.64	-	-	34 Calif. water districts
Heady et al.	-0.37	-0.17	-0.56	17 US western states
Shumway et al.	-	-0.56	-2.32	Calif. (2-eqn. model)
Shumway et al.	-	-0.48	-2.03	same (1 eqn. model)
Howitt/Watson/ Adam	-0.97	-	-	Calif. (linear programming approach)
Howitt/Watson/ Adam	-1.50	-	-	Calif. (quadratic programming approach)
II. AUSTRALIA, 1964				
Flinn				
Total seasonal demand	-0.46	-0.09/-0.25	-0.91/-1.73	5 representative farms
Spring only	-0.70	-0.09/-0.26	-1.61	in Yanco Irrigation
Summer only	0.06	-0.01/-0.03	-0.09	Area (linear
Autumn only	-0.68	-0.09/-0.25	-1.56	programming approach

Sources: Anderson (1983), chapter 3, for US data; Flinn (1969), for Australian data.

C. Demand systems

More familiar in developed economies, 'demand' systems, like the natural supplies considered in Chapter V, deliver water to farms at times and in quantities requested by the irrigator, often subject to a maximum authorised abstraction (perhaps per month and per week and per day, etc.). This more unconditional commitment to deliver water (as compared with systems (i) and (ii) has greater need of a charging scheme that encourages efficient use of water. As the value of the resource increases, prices based on volume, both that authorised and that actually abstracted, become more worthwhile. More accurate measurement and more complex allocation and charging schemes would be expected to be established, although the environmental consequences of any induced extra storage by farmers should be taken into account. At the margin the volume charge should reflect the marginal cost of system expansion as well as the operating expenses of water delivery, although subtle issues are raised in dividing the charge between authorised and actual abstractions (see V.7 and the end of VI.4 on this question). Sometimes closed pipe systems are distinguished as a fourth method of delivery; the same opportunities for efficient charging present themselves.

4. IRRIGATION WATER PRICING: PRESENT PRACTICES

We shall abstract from 'pure private property' systems of water rights in which the 'price' of the right to a certain maximum quantity of water (with associated conditions) would be established by forces of supply and demand. In extreme cases, no charging 'system' is therefore necessary.

In practice, charging systems, as can be gathered from Section VI.3, fit into one or both of two categories:

i) Flat-rate or fixed charges associated with a 'normal' allocation of water and unrelated to the volume actually abstracted (charges a function of area, crop, season, application method and/or number of irrigations);

ii) Charges related to authorised, actual or 'excess' volumes (possibly also related to season and the usual variety of block and single-rate tariffs are to be found).

Economically efficient pricing is of course only possible schemes including (ii). A study in Mexico has shown that charges based on volume taken or the number of irrigations made farmers more careful in their use of water whereas flat-rate charges gave no incentive for efficient water use (United Nations, 1980).

In Australia the price of 'public' irrigation water has in the past been relegated to a minor role; it has generally been set to cover, on an average cost basis, only part of annual servicing (maintenance and operation) costs and has often excluded completely the capital cost of structures, since huge subsidies have been granted. Yet estimates of price elasticities on representative farms have produced peak-period values consistent with the US evidence and off-peak (i.e., mid-year) elasticities in the range zero to -0.1 (see Table 11). So prices do affect demands significantly in peak irrigation seasons (spring and autumn).

The inefficiency is such that enormous investments in irrigation facilities have been found to provide little or no overall net return. Thus in Nothern Australia, the Ord River irrigation scheme has been partly developed for $A 100 million for no return at all in additional Gross National Product.

More flexible administrative systems in which the basic right to, or allocation of, water may be transferred (sold) would greatly increase efficiency by directing water to its highest-value uses. At present there are two experiments in transferring irrigation rights. In New South Wales, the annual water "allocation" (and not the "right") was recently made transferable for a one year trial for all irrigators; and in South Australia, rights have been transferable (saleable) in private irrigation areas only, subject to certain conditions. The theory is familiar: that more efficient users will be induced to purchase entitlements from less efficient users and thus bring about greater flexibility in crop and water management.

In England and Wales irrigation abstractions receive no subsidy and are subject to an almost identical charging scheme to that described in Section V.5(ii). The only difference is that spray irrigators in some Water Authorities have a two-part tariff whereby a proportion of the charge is related to the actual quantity abstracted and therefore less than 100 per cent to the authorised quantity. Usually the proportions are 75 per cent : 25 per cent.

Evidence from Greece confirms the practice of basing prices on the size of the area to be irrigated. In the Greek case, the proceeds cover only the administrative costs of the irrigation network. Permits for agricultural abstractions do not seem to be invariably required; thus in Germany some States require permits for agricultural use and some do not. If a permit is necessary, metering is required to monitor quantities but not, it seems to facilitate charging.

In Canada, too, the area to be irrigated forms the basis of the two-stage tariff structure. In Alberta, farmers pay a one-time charge of up to $50/acre as a contribution to capital costs and annual charges of $1.50 to $10.00 per acre for operating expenses. Apparently such charges imply large subsidies to irrigators for they are only meant to cover 14 per cent of major capital works (and all operating expenses). Certainly the Alberta provincial government believes irrigation must be subsidized "if it is to survive". A proposal for farmers' contributions to rise to 25 per cent a few years ago was deferred until 1985.

In the United States the dominance of the historic basis of costs is shown by the fact that prices for delivered irrigation water in California range from less than $2 to more than $200 per acre-foot. Such differences reflect, as well as the date of completion of the system, topography, type of ownership and extent of subsidy. In Japan there appear to be no charges or taxes for irrigation water.

In France water for irrigation is generally sold on the 'binomial tariff' basis outlined at the beginning of this section. However, the outstanding example of a rational conception for the pricing of irrigation water is the charging scheme administered since about 1970 by the Société du Canal de Provence et d'Aménagement de la Région Provençale, which supplies 60 000 hectares of farmland and nearly 120 communes (Jean, 1980). This scheme is thoroughly grounded in the theory of marginal cost pricing, with full recognition of the need to consider and reflect long-run costs if farmers are to make 'correct' investment decisions in terms of land, cultivation, crops, irrigation equipment and storage (and, indeed, if the Société itself is to do the same with respect to new water resource works). A peak period is identified lasting for four months from mid-May to mid-September and that plays a central role in the tariff. Tariff design starts from the objective that tariffs should reflect

-- In the peak period, long-run marginal capital costs augmented by operating costs, and

-- In the off-peak period, operating costs only.

For various practical reasons this objective sometimes has to be compromised. Thus while irrigators 'correctly' contribute to the capital costs of their distribution network only through an annual charge based upon the peak demand subscribed for by the user (the authorised abstraction), the development costs of the "main works", which in theory should similarly fall only in irrigators' demands in peak periods, are in fact lumped in with operating costs in order to establish a single year-round volumetric charge. This results from "considering the nature of the consumption which is distributed in a well-known way between the peak periods and the off-peak periods, ... [it] simplifies the price scale and enables the use of a single meter reading" (Jean, 1980).

For other sectors (urban, industrial, miscellaneous), operating costs constitute the only element in the off-peak volume charge. Off-peak demand thus has no role in determining responsibility for capital cost recovery, precisely as the economic theory of peak demand indicates (e.g., Rees, 1984). Three 'zones' are specified for the application of the tariff, further details of which are shown in Table 12.

Is it then an economist's dream? Up to a point, yes. However, the French government maintains that the agriculturalists relying on the canal are a "profession bénéficient" and therefore a state subsidy of 50 per cent on all elements of the irrigation tariff is granted. At a stroke, the price signals are clouded and the messages altered. Nevertheless, the conception remains intact.

5. THE USE OF WASTEWATER FOR IRRIGATION

Economic pricing of direct abstractions and the local availability of partially-treated sewage effluent may commend the use of effluents for irrigation purposes. The two clear advantages are:

i) The quality of surface waters otherwise receiving the effluent will be protected since even well-treated waste water may lead to secondary pollution of rivers;

ii) The use of sewage pollutants as nutrients for crops reduces the need for artificial fertilizers.

Clearly the major disadvantages are:

Table 12

CANAL DE PROVENCE TARIFF STRUCTURE

I. Financing of the major cost items

	Development cost of main works	Development cost of the distribution network	Proportional cost
Urban and industrial uses	50 % on peak flow charges 50 % on flow m³ volume at peak	No network	applies to m³ peak and off peak
Miscellaneous uses and irrigation	100 % on m³ at peak flow	100 % on flow premium	
Observations	dependent on areas	independent of areas	independant of areas when no pumping takes place

II. Details of prices for the major user

	Peak flow charge	price per m³	
		Peak	off peak
Urban uses	Identical for the two uses Varies from 1 to 8 according to areas	Varies from 1 to 2.5 according to areas	Identical for all uses and all zones
Industry	Modulated by a use/coefficient of works which is function of consumption	Slightly more expensive for industry than for urban uses	
Miscellaneous uses	Identical for all areas 6 flow categories corresponding to 0.5 - 0.8 - 1.3 - 2 - 3.1 - and 5 l/s	Varies according to areas from 1 to 6	Identical for all areas
Irrigation	Identical for all areas 7 flow categories 3.6 - 7.5 - 15 - 30 - 50 - 75 and 100 m³/h	In practice to facilitate one single reading annually single price integrating off-peak price (identical for all areas) and peak price (varying from one area to the other) On the whole price varies from one area to the other	

Farmers:

-- 50 % tariff abatement for miscellaneous and irrigation

-- minimum annual consumption of 80 m³ by m³/u subscribed to for irrigation.

Source: Jean (1980).

iii) The possible environmental and health hazards that arise, these being the more serious the less treatment the sewage has received, and also

iv) Polluted runoff from the land may produce problems of water quality.

A number of examples of the use of wastewater in this way have been documented in the Federal Republic of Germany (Fleet, 1979), including the existence of two sewage Utilization Associations (in Braunschweig and Wolfsburg) which irrigate the waste water from a population equivalent of 425 000 after only primary treatment.

A recent IWSA paper (Young, 1976) drew attention to further examples of the practice in Bulgaria, Libya, Israel and the United Kingdom. It seems to be the case, however, that, given the present state of knowledge, in most countries attempts to use effluent without considerable dilution by 'clean' river or underground water still tend to be viewed with concern.

Chapter VII

DIRECT DISCHARGES

1. INTRODUCTION

This part of the report will not be concerned with rehearsing in detail the well-known arguments about 'standards versus charges' in the regulation of direct effluent discharges to rivers and aquifers. Sufficient discussion at an international level can be found in a number of recent documents (e.g., OECD, 1977; OECD, 1980; United Nations, 1980).

It will be accepted that there are strong arguments for charges based on the marginal costs of treatment or damage and relating to the flow of pollution discharged:

i) They provide a continuing incentive for movement towards optimal discharges;

ii) Since in a given zone the charge for a given pollutant is the same for all dischargers, pollution reduction to obtain a given environmental quality objective can be achieved at minimum cost;

iii) It is a firm expression of the almost universally-acceptable Polluter-Pays Principle.

Equally, reasons may be given for preferring standards:

i) Some wastes should be avoided, perhaps above a certain concentration, at all costs. Even very high charges are too risky, and standards plus vigorous enforcement are therefore required;

ii) Historic rights to discharge are best preserved through effluent permits (assuming preservation is for some reason essential);

iii) If the major objective is to improve the quality of a stream quickly then effluent permits or the granting of subsidies tied to pollution control may do the job better than charges.

The best approach may be to combine both permits (incorporating standards) and charges into one system. But in the rest of this part we shall be solely concerned with the theory and practice of charging systems.

98

2. OBJECTIVES OF CHARGING SCHEMES

i) A scheme may attempt to alter the level of pollution in discharges towards an economic and environmental optimum. Interpreting 'economic' broadly this means that:

   a) The charge should ideally be set at <u>either</u> the estimated marginal cost of the human, biological and environmental damage (9) produced by an extra 'unit of pollution' <u>or</u> the marginal cost of augmenting and operating works (e.g., increasing the assimilative capacity of a river) to remove the possibility of that additional damage, whichever is the least;

   b) The discharging firm, if it is 'rational', will take its own effluent treatment facilities to the point where the marginal cost of treatment is equal to the charge for the marginal unit of pollution.

ii) A scheme may in some sense try to treat all dischargers equitably. This may or may not mean that they are treated equally.

iii) It may be the intention that a scheme should raise sufficient revenue to compensate pollution-sufferers or to finance a programme of anti-pollution works (note, however, that redistribution of charges is not necessary for economic efficiency; indeed, it may give rise to 'excessive' expenditure on pollution control).

Granted these objectives, it is possible to assess a given charging scheme by its effects on:

i) Economic efficiency;

ii) Equity;

iii) Income-generation (Thackray, 1976).

3. PRESENT CHARGING SCHEMES

A number of European countries have recently introduced water pollution charges (e.g., France, 1968; Netherlands, 1969). Parts of the Federal Republic of Germany had schemes originating in the early years of this century, but a Federal Act passed in 1976 laid down that the States should by 1981 levy a uniform national system of wastewater charges on the direct discharge of specified effluents into public waters.

The authors of the French scheme accepted the premises of a marginal approach and thus the validity of the efficiency objective. It was claimed that the aim of the scheme ".....is ultimately to equate private costs (the payment of charges or expenditure on water management) with social costs (damage prevented or compensated)" (OECD, 1980, p. 80). But in 1980 the then present charges were not being set at optimum economic rates: "the rate will have to be appreciably increased if the charges are to be effective and [act as] incentives." In the Netherlands "the levies are changed so as to be in a position to take measures to combat pollution and are therefore imposed for specific purposes" (i.e., pollution control works and equipment). Consequently they do not reflect directly the damage caused by the particular discharge.

In Germany the basis of the 1976 Act was economic:

"The Waste Water Charges Act is the first environmental protection act [in the Federal Republic] which is based upon the "polluter-pays" principle using charges as an economic tool .... the term "economic tool" means that the most favourable action alternative can be chosen by a particular company by way of deciding either to pay duties or to implement measures to avoid pollution; ..... formerly "external" costs (i.e., costs which are not taken into account in the polluter's costing) are transformed into internal costs as far as possible" (Federal Ministry of the Interior, 1983).

The charges are levied "per unit of noxiousness", with five groups of pollutants translated into units of noxiousness: settleable solids, COD, cadmium, mercury and toxicity for fish. The Federal government, with the aid of 60 task forces, has laid down minimum discharge standards compatible with 'commonly accepted technology' for individual industries and municipalities of different sizes. If direct dischargers meet these minima or any more stringent standards that individual States may lay down, the charges payable are reduced by 50 per cent. Revenue from the charges must be used for administering and enforcing the new scheme and also "... for specific purposes connected with measures for maintaining or improving water quality" (Article 13, Waste Water Charges Act, 1976). Possible exemption from the obligation to pay charges in cases of hardship has been provided for (see section X.4, below).

In the Netherlands as well as Germany there was much uniformity among rates of charge (in the case of the Netherlands, along individual rivers), but in France a much more complex system exists which, at least in principle, should give greater allocative efficiency. The French scheme in Seine-Normandie has the effect of encouraging industrial development in the lower reaches of the Seine and in the estuary (i.e., where it exists already) and not encouraging it (by making pollution charges higher) in coastal bathing areas and along stretches of rivers governed by national decrees on quality objectives (OECD, 1980).

All these countries have experienced or anticipate significant increases in the real value of charges. In (OECD, 1980) the 'real' levies applicable along the Dommel and Aa rivers were forecast to increase by 22 per cent and 29 per cent respectively over the 1975-85 decade, having already risen by about 50 per cent and 150 per cent over 1971-74. In the Seine-Normandie Agence de Bassin real charges rose 82 per cent over the "1st and 2nd programmes" of 1969-75; and the Waste Water Charges Act in Germany has had charges increasing from DM 12 to DM 40 per unit of noxiousness over the 1981-86 period: a nominal increase of 233 per cent. Despite past increases effluent fees remain small compared to the total cost of waste treatment (including Sewage Treatment Works). In France the per capita effluent charge has been estimated as less than 10 per cent of the total per capita cost of secondary treatment services for an average city. The corresponding proportion in the Ruhr is about 30 per cent (United Nations, 1980).

Revenue from charges covers 100 per cent of pollution control expenditures along the Dommel and Aa basins, and about 70 per cent of the amount allocated by the Seine-Normandie Agence for assorted anti-pollution projects in the 1976-81 action programme.

4. EFFECTS OF CHARGING SCHEMES

Few empirical results have yet been reported. In the Netherlands in many cases companies were induced to decreasing emission of waterborne pollutants and switched to purifying the effluent on their own account (OECD, 1980); one Dutch author noted that industrial water consumption had decreased by 30 per cent while industrial production grew over 1970-76 and claimed that the water-conserving measures responsible had arised in part from the 1969 Pollution of Surface Waters Act (which introduced charges) (Snijders, 1980). However, the United Nations study issues a caution:

"The fact that the introduction of a charge system is followed by a major investment in treatment systems, or by a decline in pollution levels, does not establish any firm conclusions about the efficiency of the charge system. In every case where this phenomenon has occurred, other policy intruments accompanied the introduction of the charge system, and it is not possible to separate the cause and effects of each element." (United Nations, 1980, p. 138).

It followed that for France

"Perhaps accelerated investents in treatment facilities.....were due to the subsidy system introduced at the same time." (ibid)

On the other hand, the authors of the official report on the first five years' operation of the German Waste Water Act (1978-82) seemed in no doubt that the mere passing of the Act had created significant responses. Although the interpretation by the States of the Act has varied, many municipalities had built sewage treatment plants that went beyond the requirements of the law; 20 per cent had speeded up building and 30 per cent had altered plans even before they were obliged to. Between 1977 and 1979 more than half of the industrial direct polluters took measures to improve their waste water treatment processes and/or built the appropriate plants. In early 1980 a further improvement in the technology of the treatment plants of 66% of the direct discharging firms was observed. By 1981 more than a half of waste dischargers were meeting the Federal minimum requirements. There are also reports from the pulp, electroplating and textile refinement industries of other innovative reactions to the Act which induced a general modernization of production processes and therefore enhanced competitiveness as well as large energy savings. Economy washing processes and materials reclamation were frequent examples. Two major problems have arisen: the number of dischargers that have avoided payment, and charges for the discharge of rainwater. On these issues, differences between the Lander have given cause for concern. (Federal Ministry of the Interior, 1983; Brown and Johnson, 1984).

5. EVALUATION OF EFFLUENT CHARGE POLICIES

The United Nations recently evaluated five European charging policies on efficiency grounds by selecting four criteria and running tests on each policy. The results are reproduced in Table 13, the 'correct' answers to (a)-(d) being 'Yes' in each case.

No definitive conclusions follow since, for example, although France has 3 x Yes as compared to 2 x Yes for the Ruhr, the Polluter-Pays Principle is more closely upheld in the Ruhr because the level of charge is higher and the rate of subsidy is lower.

6. DESIDERATA OF AN EFFLUENT CHARGING SYSTEM

The authors of the United Nations study, having undertaken the European evaluation, listed six major considerations for guiding the establishment of a practical effluent charging system. It is useful to quote their suggestions here:

"a) When an effluent charge system is introduced, initially low rates may be established, with a clear indication of the amount and timing of rate increases. This is illustrated by the experience in France, Hungary and the Netherlands. A graduated charge system at the time of implementation decreases the cost of adjusting to the new circumstances;

b) The charges should be based on pollution coefficients that are comparatively easy to measure by techniques that yield consistent results. Effluent permits can be relied on to protect the water bodies from quality degradation due to pollutants which do not enter in the determination of the effluent charge;

c) The table of pollution coefficients establishing levels of pollution per unit of output (input) or per population equivalent for the industrial sectors and municipalities greatly simplifies the administration of an effluent charge system. In order to maintain individual incentives to reduce pollution, the table of pollution coefficients must be revised periodically, and there must be a reasonable provision for either the managing agency or the polluter to sample effluents and establish payments according to the actual damage caused by the discharge of pollutants;

d) Effluent charges will perform best in an institutional structure that encourages co-operation to achieve economies of size from large-scale waste-treatment facilities. This seems to be best exemplified by the water quality management association in the Ruhr

Table 13

EFFICIENCY ASPECTS OF EUROPEAN EFFLUENT CHARGE POLICIES

Criterion	France	Federal Republic of Germany (Ruhr)	Hungary	Netherlands	
				National	Waterboards
(a) Non-uniform treatment of firms in same industry producing different levels of waste	No	No	Yes	No	Yes
(b) Locational distinctions recognised	Yes	No	No	No	Yes
(c) No economically arbitrary discrimination by size	Yes	Yes	Yes	Yes	?
(d) Intertemporal variation acknowledged	Yes	Yes	No	Yes	Yes

Source: United Nations, 1980, p. 141.

region of the Federal Republic of Germany. It is also well illustrated by the pronounced trend towards comprehensive management in Western Europe, which features the consolidation of water organisations;

e) An effluent charge system should emphasize regional differences. It is very likely that charges should be set high for polluters located among streams that are subsequently used for potable water or swimming. There may also be rivers for which the marginal value of added water quality is significantly lower, and charges can be set relatively lower to reflect this fact;

f) As a practical matter, standards can be used to ensure that water quality is maintained in a certain acceptable range of predetermined levels and that charges can be varied until this range is achieved. Moreover, effluent charges should be established in such a way that they can become more differentiated as time passes."

(United Nations, 1980, p. 142)

Chapter VIII

THE METERING DECISION

The discussion in this part of the report will be focussed on domestic metering, although the analysis of the "metering decision" is the same whether we are concerned with farmers in Avignon, commercial offices in Amsterdam or householders in Auckland.

1. PRESENT VARIATION IN PRACTICE

Some indication of the enormous variation in practice in Member states in respect of domestic metering will have been gathered from Section III.4, above. Table 14 amplifies and underlines that variation.

Clearly it is in Britain that domestic metering is least prevalent (but see VIII.4, below, on current policy). Only in Norway is metering also the exception rather than the rule. At the other end of the scale, in many states metering is very comprehensive (e.g., in France, Finland, Japan and Switzerland). Possibly France charges a higher proportion of individual households by volume than any other country. In most countries practice on metering individual households in apartments seems to vary greatly. Thus Coe reported some years ago that whereas the private water company serving Barcelona had a policy on installing (and charging by) a separate meter in the basement of an apartment block for each consumer in the building, in Madrid a master-meter was placed by the local undertaking, with individual consumers sharing out the charge by various means of their own choosing, including the installation of individual privately-owned sub-meters (Coe, 1978).

The mixture in Australia was, at least until recently, rather more bizarre than elsewhere. As in Brisbane, high minimum rates can occasionally transform a rateable value system into a poll tax on residences (a fixed charge per property for the majority of residents) while in Melbourne, as has already been noted (Chapter III.4), the presence and operation of meters and meter-readers are of only limited use since most households stay within a 'free block' and pay just the rateable value component of a theoretically two-part tariff.

Two countries have reported examples of optional metering, i.e., optional at the request of the householder (usually there is an installation cost and meter rental involved). In England & Wales optional

Table 14

DOMESTIC METERING IN OECD MEMBER STATES*

Country	Extent of Household Metering	Comments
Australia	All metered but large 'allowances' mean that only about x % of households pay for any water by volume	
Canada	60 % of utilities surveyed in 1983 used a flat-rate tariff	
Denmark	In towns virtually all metered; in rural areas usually unmetered	
Federal Republic of Germany	Only 30 %-40 % of households directly metered (very few in individual apartments)	
Finland	Non-apartment households metered? Approx. position in apartments?	In apartments, separate water/ wastewater charge usually based on no. of persons living in flat.
France	All separate dwellings metered; individual apartments: 50 % metered for cold water and 85 % for hot (1972 figures)	meters read twice a year
Japan	All households metered (including those in apartments)?	Meters in households read every one or two months.
Netherlands	75 % separately metered 6 % collectively metered 19 % unmetered	Local water works company discretion exists 17 out of 100 cos. reported 100 % metering (in 1982, 32 % of Dutch dwellings were flats. Thus, 81 % of flats had separate meters?)
Norway	About 10 %?	Local discretion; 100 % metering of new dwellings in some areas. Some 'new' metering of old properties.
Switzerland	90 % of all buildings supplied by public water supply have meters	
U.K. (England & Wales only)	0.36 % of households have meters	At discretion of individual house-hold, except compulsory metering of all owners of automatic lawn sprinklers in South-West WA (April 1984) and Welsh WA (April 1985)
United States	96 % of connections metered in 132 large utilities serving 70 m. people in total; approx. 25 %/30 % of dwellings collectively metered and 5 % unmetered	Calculated from 1981 A.W.W.A. survey and 1982 Census. Usual billing period: every 2 or 3 months.

Sources: Country submissions.

*   Individual households assumed to pay (at least in part) by volume if metered, unless indication to the contrary.

    Hot water provided in apartments under district heating schemes is normally metered.

105

metering began to be offered by the water authorities and private water companies in the early 1980s, but the choice has not proved very attractive. By early 1985 only 1/2 per cent of 18.4 million householders had opted for a meter, the figure being highest in the Anglian Water Authority (about 2 per cent). In the Antwerp undertaking, in Belgium, the fixed tariff system (see III.4, above) attracts 80 per cent of households; the other 20 per cent are either obliged to install a meter, because of the presence of a water-intensive appliance, or they choose to do so.

2. THE METERING DECISION

The metering decision has two dimensions, equity and efficiency. The equity argument for metering is clear; it permits volumetric charging, which means payment according to the quantity of water used (and, approximately, the quantity disposed of into the sewers). The efficiency dimension is best approached by considering metering like any other investment decision, as being characterised by the accumulation of benefits and the incurring of costs. A fundamental distinction must, however, be made between economic and financial appraisal.

A. Economic Appraisal

Economists commend cost-benefit analysis to evaluate the desirability on efficiency grounds of changing all or some domestic water consumers over to a metering and volumetric charging system.

It is shown in Annex 3 that domestic metering and unit quantity pricing is to be recommended if economic gains outweigh economic losses, i.e., if

$$\sum_{i=1}^{n} C_i \cdot R_i + I > M + U$$

where C_i = long-run marginal cost of water supply (and sewerage and sewage disposal) provision in peak or off-peak period (or season) i,

R_i = reduction in water supply (or sewerage and sewage disposal) demands in peak or off-peak period (or season) i,

I = benefits of improved waste reduction and demand forecasting (each year),

M = direct resource costs of metering and volumetric charging (each year),

U = value of water consumption foregone, (each year),

and n = the number of periods or seasons into which the year is divided.

The United Kingdom National Water Council undertook a cost-benefit assessment of domestic metering in 1976 (National Water Council, 1976). Using a framework similar to that just outlined, it assumed $I = 0$ (I is difficult to value) and also $U = 0$, on the implicit grounds that the loss in value of water forgone (U) was just balanced by the gain in equity among water consumers.

The question thus reduced to:

$$\text{is } C.R > M ?$$

$$\text{i.e. is } C > M/R ?$$

All costs (C_i) and water service demand reductions (R_i) were encapsulated into single terms, C and R. C was estimated as 21 pence per cubic metre from cost data generated by a national water resources expansion programme conceived a few years earlier. It was more convenient to estimate M and R in present worth terms, and M/R was calculated at 23p./cubic metre. R was assumed to be 20 per cent of pre-metering demand.

$$\text{Therefore, } C < M/R$$

Thus it was concluded that universal metering would be unjustified on economic grounds, although selective metering where installation costs were low and marginal costs high could lead to efficiency gains.

Hanke (1982) used a "short-cut" procedure for a cost-benefit decision on the extension of metering in Perth, Australia, by assuming,

a) $n = 2$,
b) $i = 1$ corresponding to summer use
         and $i = 2$ corresponding to winter use
c) $U = 1/2 P.R$, where P = price of water after metering, and
d) I cannot be valued.

The metering decision thus reduces to

$$\text{is } C.R > M + 1/2 P.R ? \quad \text{(again I cannot be valued)}$$

$$\text{i.e. is } R(C - 1/2 P) - M > 0 ?$$

With the separate summer and winter calculations for R and C, Hanke's estimates are as shown in Table 15.

The figures are aggregates and apply to the 17,986 remaining unmetered water connections who used an estimated 14 per cent of Perth's water production in 1976-77. Savings due to metering were assumed to be 30 per cent. The conclusion is that the extension of metering would have been economic.

In a much more detailed inquiry in the United Kingdom, a Department of the Environment study of domestic water metering (1985),

a) Recognised and incorporated the different long-run marginal costs of average day, peak-week and peak-hour demands,

b) Postulated differentiated reductions, following metering, in in-house and ex-house water use (assumed to be 12 per cent and 30 per cent respectively), thus giving significantly higher savings in peak-week and peak-hour consumption, and

107

Table 15

ANNUAL BENEFITS AND COSTS OF WATER METERS IN PERTH, AUSTRALIA

	R (C-1/2P)	less	M	=	?
Summer :	5.6 m.m^3 ($0.176/m^3-$0.053/m^3)				
Winter :	2.2 m.m^3 ($0.096/m^3-$0.053/m^3)				
Total :	$783 400	- $241 342		=	$542 058

m.m^3 = millions of cubic metres

Source: Hanke (1982), Table 1.

c) included (in I) an estimate of the value of reduced leakage in the consumer's supply pipe if metering was to be introduced.

Estimates of C were calculated as 20 pence/m^3 and 26p./m^3 in two water authorities, while M/R less I/R was estimated as 27p./m^3. Thus universal metering (or, what amounts to the same thing, metering the average consumer) did not appear to be economically viable. However, it was stressed that many of the assumptions made were uncertain and that further work -- both field trials and economic investigation -- was essential. Nevertheless, for the smallest 20 per cent of water users -- those households that use less than half the average -- the inquiry concluded that there was no prospect of viable metering.

B. Financial Appraisal

It may also be desirable to appraise the desirability of domestic metering to the water utility on financial grounds. This approach assesses the balance in the changes in cash inflows and outflows that would result from metering, and would be necessary for calculating the consequences for consumers' charges. Financial appraisal is also considered in Annex 3.

Rotterdam water undertaking costed out the adoption of domestic metering throughout the city in 1968 and estimated that the "critical cost" element of M came to 135 Fl (about £18) per supply. The idea was dropped on cost grounds and a few years later had not been reconsidered (Coe, 1978). Another author, perhaps referring to the same calculations, reported a necessary increase of 35 per cent in the price of metered water in the city if domestic metering was to be financially viable (Snijders, 1980).

3. THE IMPACT EFFECT OF METERING

This important aspect of the metering decision may be judged in two ways: by examining the experience of those communities that have been switched over from a non-metered to a metered unit quantity charging situation, and by comparing the experience of metered and unmetered communities at approximately the same time. Comparisons generally reflect differences in waste as well as those in residential consumption.

Table 16 shows that on both approaches sizeable savings due to metering are revealed. The higher percentage savings in peak consumptions suggest particularly large reductions in external-to-the-house use. The writer is not aware of a single serious study that, taking account of other factors present, has not suggested that the impact effect of metering is significant, rapid and sustained.

4. PRESENT POLICIES TOWARDS METERING

The country submissions have suggested that in a few countries attitudes and policies towards domestic metering are in a state of flux.

In the United Kingdom a large study of the subject was initiated by the government in October 1984. It seems to have been prompted by

a) Higher-than-expected peak demands in the relatively hot summers of 1983 and 1984;

b) Recent advances in customers telemetry possibilities (see III.5, above); and

c) The greater 'commercialisation' of the water industry desired by the government (see II.2, above).

Results were published late in 1985 (Department of the Environment, 1985), as described in VIII.2(A). The Steering Group responsible for the study felt that the more complex economic analysis used needed to be tested "by live, obligatory trials, which will not only test basic assumptions, but will also evaluate the economy of bulk installation and the effect on demand of different tariff structures." The trials are expected to extend over three years or more.

In Australia, as explained above, hitherto there have often been very large 'allowances' for domestic consumers, which have meant that, despite meters being installed in most dwellings, the price of marginal litres has been zero for nearly all customers. A number of utilities have recently drastically reduced the allowances (e.g., Perth, Hunter District, Toowoomba) and current policy is to encourage the adoption of more rational water pricing policies for all sectors.

Table 16

ESTIMATED SAVINGS DUE TO CHARGING BY VOLUME

I. CROSS SECTION STUDIES

Location	Comparison	Savings due to metering		Reference
Gothenburg, Sweden	Single houses metered (M) with apartments unmetered (UM)		33 %	Shipman (1978)
Copenhagen, Denmark	63 metered systems and 148 non- or part-metered systems		20 %	Shipman (1978)
Severn-Trent Water Authority, UK	Domestic consumers in Malvern (M) and Mansfield (UM)		10 %	Thackray, Cocker & Archibald (1978)
Toronto, Canada	?	winter : summer : year-round :	32 % 16 % 25 %	United Nations (1980)
United States	10 metered and 8 unmetered areas (domestic consumption only)	year-round : max. day : peak hour :	34 % 58 % 52 %	United Nations (1980)
Towns in Alberta, Canada	Per capita consumption in towns: M towns & UM towns in early 1970s		35 %-40 %	Gysi (1981)
Alberta, Canada	Calgary (80 % of domestic sector unmetered); and Edmonton (100 % unmetered); both cities about 100 000 residential customers over 1972-75	year-round :	52 %-57 %	Gysi & Lamb (1977)
	Calgary: metered and unmetered households over 1972-75	(1974) year-round : (1975) peak month : peak month :	45 %-55 % 66 % 72 %	
Denver, USA	M and UM households		19 %	Environmental Defense Fund (1980)

Table 16 (cont.)

II. TIME SERIES STUDIES

Location	Year(s)	Average Consumption		Saving	Reference
		Before metering	After metering		
Fylde, England (291 properties in trial)	1970-71 to 1971-72	293.0 lpd	261.0 lpd	11 %	Jenking (1973)
Kingston, N.Y., USA	1957-63	20.7 Mld	15.4 Mld	26 % (a)	Herrington & Tate (1971)
Denver, USA	1957	691.0 lhd	639.0 lhd	7 % (a)	Snipman (1978)
Philadelphia, USA	1955-62	2 562.0 lpd	1 408.0 lpd	45 % (a)	Herrington & Tate (1971)
Zanesville, Ohio, USA	1958-61	632.0 Mld	492.0 Mld	22 % (a)	Herrington & Tate (1971)
Boulder, Colorado, USA	1961-63	1 160.0 lhd	748.0 lhd	36 % (a)	United Nations (1980)
Fredericia, Denmark (109 1-family houses)	1971 (Winter) -1972 (Summer)	1 010.0 lpd 1 470.0 lpd	460.0 lpd 830.0 lpd	54 % 44 %	
6 small communities (4 232 households) region of Durham, Canada	1978-80		various	13-20 % (a)	Loudon (1984)
Moss City, Norway non-industrial use: Total use:	1979-83 aver.day 1978-83 peak day	356.0 lhd 880.0 lhd	323.0 lhd 520.0 lhd	9 % (a) 41 % (a)	(Country) (submission)
Asunto Oy Unioninrinne, Helsinki, Finland (37 flats; separate metering of hot water)	1980	66.0 lhd	63.0 lhd	5 %	Finnish Energy Economy Association (1983)
Toowoomba (b), Queensland, Australia	1978-79 (aver. day) 1979 (peak day) 1979 (peak month)	450.0 lhd 864.0 lhd 680.0 lhd	350.0 lhd 680.0 lhd 399.0 lhd	22 % (a) 21 % (a) 41 % (a)	

Table 16 (cont.)

Location	Year(s)	Average Consumption Before metering	Average Consumption After metering	Saving	Reference
Hunter District (b), Australia	1982 (aver. day)	290.0 lhd	199.0 lhd	31 %	Broad (1984)
	1982 (peak day)			50 %	
	1982 (losses)	24.4 Gl	16.4 Gl	33 %	
				20 %	
Perth, Australia (b)	1975/76-1980/81	508.0 Kl	307.0 Kl	20-30 %	Metropolitan Water Authority (1985) (c)
Hot water in apartments					
Lille (350 flats)	1965-70	77.0 lhd	33.0 lhd	57 %	
Colmar (493)	1967-72	60.0 lhd	42.0 lhd	30 %	Vermersch (1974)
Roubaix (132)	1964-69	61.0 lhd	27.0 lhd	56 %	
Cold water in apartments					
Nanterre (250)	1968-73	138.0 lhd	87.0 lhd	40 %	
Brest (714)	1965-70	73.0 lhd	52.0 lhd	27 %	
Noisy (179)	1966-74	129.0 lhd	89.0 lhd	31 %	

a. 'saving' percentage includes reductions in unaccounted-for water (primarily leakage).

b. Introduction of complete metering and an 'effective' pricing policy.

c. Estimate for reduction in demand due to changes in consumer behaviour following introduction of pay-for-use pricing; other factors contributing to 20l Kl decline in per household use were reduced household size, reduced level of real income and increased bore ownership (ground water abstractions).

lhd = litres per head per day

lpd = litres per property per day

In the Netherlands, discretion concerning domestic metering is left entirely with the local undertaking. One large utility, Waprog (N.V. Waterleidingmaatschappy voor de Provincie Groningen), intends to meter the 75 per cent of its domestic consumers that are at present unmetered in 1986 and 1987. This will involve installing about 100 000 meters in 20 months. The only reason for metering in this case is the greater equity that will result; the expectation is that domestic consumption (at present 127 lhd) will decline by only about 6 per cent.

New houses have been prepared for metering at the time of their construction in Norway at least since 1972 (Coe, 1978). Many new homes are now being metered as they are built, although discretion remains with local authorities. The community of Moss (population 25 000) metered all households between 1978 and 1984, with striking results for peak-day consumption (see Table 16, II).

The metering of individual apartments is becoming more common in Finland. To obtain State assistance for residential building projects guidelines concerning the metering of water consumption must be followed (introduced in October 1982):

"Terraced houses and groups of one-family houses organised into 'company' form: meters shall be installed for the separate metering of both cold and hot water use in each dwelling.

Blocks of flats: water pipes and appliances shall be installed in such a way that it is easy to install (later) meters for the metering of water use in each flat separately.

Meters for the metering of hot water use in each flat (separately) should be installed at the construction phase if there is no good reason for later installation. The meters should be such that the remote-reading of them is possible." (National Housing Board of Finland, 1982)."

These guidelines cover about 50 per cent of residential building activity in Finland. In older housing corporations (responsible for apartment blocks), the existing company legislation often inhibits charging of individual flats for water consumption. An official proposal is now under consideration, to make it easier to amend a company's articles (e.g., to move to individual charging). Difficulties in metering and charging (by volume) individual households (administrative costs, etc.) are, however, recognised in Finland.

Chapter IX

SUBSIDIES IN WATER SERVICES

The extent of subsidies in the water industry in a number of countries has already been noted. By lowering prices below cost, the result has often been "...the over-building of systems, waste of public funds and sub-optional water use practices." (Canadian submission). In this part of the report available information from OECD countries on the present magnitude and types of subsidisation is brought together.

1. DEFINITION AND TYPES OF SUBSIDY

A subsidy may be defined as a transfer of money between i) a unit of federal, state, regional or local government and ii) a water services utility, other than a charge for goods provided or services rendered.

Generally, the transfer is from government and to the water utility. Much more rarely, however, there exist negative subsidies whereby part of a utility's surplus is collected by government, over and above any interest due on loans.

Subsidies may take two forms: grants and loans provided at artificially low rates of interest. They may be directed at two types of expenditures, capital or current (operating and maintenance expenses, plus debt interest due). Subsidisation of the first category is both more widespread and relatively larger.

2. CURRENT PRACTICE

The extent of subsidisation varies greatly both within and across OECD countries, as Tables 17 to 20 make clear. In general, it is higher outside Europe -- in North America, Japan and Australia. Information available relates almost wholly to grants; too few details on low-interest loans have been provided to permit calculation of the size of the subsidy implied.

Table 17

SUBSIDY RATES FOR NEW CAPITAL EXPENDITURES IN PUBLIC WATER SUPPLY AND SEWAGE DISPOSAL SERVICES, EARLY 1980s

	Grant from			Residual financing (usually, a water utility loan)
	Federal or central government	State, region or prefecture	Local authority	
JAPAN				
Distribution projects	25.0-50.0 %	–	–	50.0-75.0 %
Local (a) sewers: 'subsidised'	60.0 %	–	6.0 %	34.0 %
'self-financing'	–	–	10.0 %	90.0 %
Regional (a) sewers: 'subsidised'	66.7 %	8.3 %	–	25.0 %
'self-financing'	–	10.0 %	–	90.0 %
Local (a) sewage treatment works: 'subsidised'	60.0-66.7 %	–	6.0-5.0 %	34.0-28.3 %
'self-financing'	–	–	5.0 %	95.0 %
Regional (a) sewage treatment works: 'subsidised'	66.7-75.0 %	8.3-1.6 %	–	25.0-23.4 %
'self-financing'	–	10.0 %	–	90.0 %
NEW ZEALAND				
Water resources and trunk mains	30.0 %	–	–	70.0 %
SWITZERLAND				
Water supply: construction works	5.0-40.0 %			60-95 %
Water supply: (smaller waterworks)	up to 70.0 %			more than 30 %

Sources: Country submissions.

a. 43 per cent of local sewer expenditures were of the 'subsidised' category (57 per cent 'self-financed'); 90 per cent of regional sewer expenditures were 'subsidised' (10 per cent 'self-financed'); and 95 per cent of all sewage treatment works were in the 'subsidised' category (5 per cent 'self-financed').

Table 18

EX POST SUBSIDISATION OF CURRENT (Curr.) AND CAPITAL (Cap.) EXPENDITURES IN PUBLIC WATER SUPPLY (PWS) AND SEWAGE DISPOSAL (SD) SERVICES

	Federal or central government	Grant from — State, region or prefecture	Grant from — Local authority	Residual financing (loan or charges)
AUSTRALIA (early 1980s)				
PWS & SD: Curr.: larger urban centres	–	20.0 % (a)	–	100 %
PWS & SD: Cap.: smaller urban centres	–	12.5 % (b)	–	80 %
PWS & SD: Curr.: smaller urban centres	–	15.0 % (c) / 25.0 % (d)	–	475-87.5 %
CANADA (1980)				
Waterworks and waste treatment: Cap.	[small]	$0.2 billion	–	$1.1 billion
FINLAND (1982)				
PWS: Curr.	–	–		100 %
PWS: Cap.	–	–	8.0 % (e)	92 %
SD : Curr.	–	–		100 %
SD : Cap.	–	–	54.0 % (e)	46 %
FRANCE				
PWS: Curr. (1981)	15.0 %	3.0 %	15.0 %	100 %
PWS: Cap. (1981)	–	–	–	67 %
SD : Curr. (1978-80)	–	–	–	100 %
SD : Cap. (urban) (1978-80)	21.0 %	13.0 %	2.0 %	64 %
JAPAN (1981)				
PWS: Cap.		5.2 %		94.8 %
SWITZERLAND (1982)				
PWS: Cap.			24.0 % (e)	76 %
UNITED STATES (1982)				
PWS: Curr. (Govt-owned utilities)	61.0 %	3.0 %	–	97 %
SD : Cap.		9.0 %		30 %

Sources: Country submissions and Boland (1983).

a. Queensland;

b. New South Wales;

c. South Australia;

d. Tasmania;

e. Includes central and local government grants.

Table 19

EX POST SUBSIDISATION AND SUBSIDY RATES OF PUBLIC WATER SUPPLY AND SEWAGE DISPOSAL SCHEMES IN RURAL AREAS

	Grant from			Residual financing (loan or charges)
	Federal or central government	State, region or prefecture	Local authority	
I. EX POST SUBSIDISATION				
AUSTRALIA (early 1980s)				
PWS & SD: Northern Territory (Curr.)	-	70 %	-	30 %
PWS: W. Australia (Curr.)	-	41 %	-	59 %
SD : W. Australia (Curr.)	-	65 %	-	35 %
FRANCE (1978-80)				
SD : Capital Costs only	11 %	14 %	19 %	56 %
II. SUBSIDY RATES				
NEW ZEALAND (1983)				
"community rural water supply" (Cap.)	50 %	-	-	50 %

Sources: Country submissions.

Table 20

EX POST SUBSIDISATION AND SUBSIDY RATES FOR IRRIGATION PROJECTS

	Grant from			Residual financing (loan or charges)
	Federal or central government	State, region or prefecture	Local authority	
I EX POST SUBSIDISATION				
AUSTRALIA				
W. Australia (early 1980s) (Curr.)	-	71 %	-	29 %
UNITED STATES				
Many federal irrigation projects	?	?	?	10 %
CANADA, Alberta (1982; Cap.)	?	?	?	14 %
II. SUBSIDY RATES				
NEW ZEALAND (Cap.)	70 %	-	-	30 %
FRANCE (a) (Curr. Costs)	50 %	-	-	50 %
JAPAN (Curr.)	45/60 %	10/25 %	-	20/45 %
CANADA (Cap.)	50 %		50 %	-
UNITED STATES (Bureau of Reclamation projects: curr. costs)	-	60 % (b)	-	40 %

Sources: Country submissions; United Nations (1980).

a. 50 per cent subsidy to farmers (as a "profession beneficient") if water authority is developing water resources in a "water-poor" area (e.g., Canal du Provence)

b. Subsidy mostly paid by other users of project, especially power.

With few exceptions, current expenditures on public water supply systems (operations, maintenance and debt interest) are covered by user charges. This is also true for a number of countries not listed in the tables (e.g., England & Wales, Netherlands). The reported exceptions are in smaller urban centres in Australia (up to 25 per cent state funding of operating costs), in Germany (where "... the allocation of cost to an individual [public] service can....be very difficult.... therefore direct and indirect subsidies are quite common with regard to the public water supply), in Canada (where industries are reported to frequently pay prices below cost, for reasons to do with the promotion of a municipality or larger area) and in the United States (where 10 per cent of non-private utilities were receiving subsidies from municipalities in 1982, amounting in total to about 3 per cent of total revenues).

A number of countries reported instances (seemingly infrequent) of water charges being used to finance other local services (negative subsidies). In Canada there are cases of water service revenues being placed in general municipal funds; similary in Germany profits from water are occasionally used to cover losses in other services. The phenomenon is "rarely" observed in the Netherlands, where in any case "improper" profits would be moderate (less than 3 per cent of total receipts).

Additionally, we should note the cross-subsidisation across generations of consumers which may arise from historic cost accounting and the associated historic cost depreciation. The consequence of this practice is that often there are only small amounts of funds in reserve for new works and deferred maintenance, so that charges have to be increased dramatically when a new works is completed. The result is that at times of steady and sustained inflation the generation of consumers who first realise the benefits of new capital works effectively subsidise both preceding and following generations.

Subsidisation of capital expenditures is much more frequent and also greater, as Tables 17 and 18 show. Even in Europe it can be large: witness France (where, however, 93 per cent of State grants in 1981 went to rural supply schemes) and Switzerland (one-quarter of 1982 investment from subsidies). Indeed, in Switzerland, up to 70 per cent of the new capital of smaller waterworks may be provided by the federal or regional authorities or by fire insurance companies, who have a long history of financial support for water supply in Switzerland arising from their direct interest in a good and reliable supply for fire-fighting.

In wastewater services, subsidies are in general higher. The proportion of capital costs having to be found through loans -- and therefore serviced by charges levied on users -- was only 30 per cent in the United States (in 1982), 46 per cent in Finland (1982), 35 per cent in Western Australia and 64 per cent in France (end-1970s).

Rural area subsidies for both water supply and sewage treatment services seem to be higher still, on the limited evidence available (see Table 19). Other countries not listed report special help for rural districts, e.g., in Western Canada the Prairie Farm Rehabilitation Administration, a Federal agency, assists in the financing and construction of farm water supplies and small community systems, and in the United Kingdom "small government grants" are available for first-time rural water supply and sewerage services.

119

The highest subsidies are, however, found for irrigation projects (see Table 20); here subsidy rates in the 50 per cent to 80 per cent bracket are common. Huge subsidies have, for example, been granted in Australia, where, in one recent study (of the Bundaburg scheme) charges were found to amount to only 33 per cent of overall economic costs (for supply by channel) and 10 per cent (for river supply). Systems have consequently been over-built and agricultural resources misallocated.

Finally, a number of submissions have reported or implied extensive cross-subsidisation across geographic areas and across different groups of users of a particular water service (see II.2, above). For example, Adelaide metropolitan area consumers in South Australia heavily subsidise rural areas in that state. Quantification of such cross-subsidisation is not available.

3. CHANGES IN SUBSIDY POLICY

Subsidy policy and practices are not, however, static. Both the Australian and Canadian submissions are highly critical of subsidies because of the consequential inefficiency. Initiatives in Australia reported in this paper (pay-for-use pricing, freeing of water rights) suggest a rapid change of policy emphasis in that country. No such changes have been reported for Canada.

In the United States, moves to reduce subsidies are in evidence. Before 1972, recipients of federal aid for wastewater projects were simply required to provide for proper operation and maintenance. There was no specification of how these activities should be financed. From 1972 onwards, municipalities were required to adopt a user charge system to recover operation and maintenance costs; otherwise, no federal grant. Finally, in 1981, new regulations stressed the importance of a user charge system that would produce revenues adequate also for the replacement of the system. Since 1981 the scope of the programme has been reduced as part of federal cost-cutting, and Department of Agriculture grants for water supply system construction in rural or semi-rural areas have been phased out completely.

In France the large state grants for capital investment (Table 18) have now been replaced by block grants from the central government, which local authorities may expend as they wish on various local services. Results of this new freedom are not yet available.

In England & Wales water authorities have been virtually self-supporting since the major reorganisation of 1974. The only sign of movement in the opposite direction, giving increased scope for subsidisation, is from Denmark, where the Water Price Committee (National Agency of Environmental Protection, 1984) recommended a new section in the existing national Water Supply Act so as to reduce restrictions on local authorities offering financial support to private common water supply plants. In "special" conditions, local authorities would be given power by this legislation to subsidise operations expenditures, although in return the local authority would be able to impose conditions, for example with respect to the design of tariffs and accounts.

Chapter X

MOVING TO A MORE RATIONAL PRICING SYSTEM

1. INTRODUCTION

Limited evidence has been accumulated concerning particular economic and environmental problems which may arise from the introduction of new or significantly amended water charging schemes. Chapter X.2 sets out the 'economic' effects that might occur; they are associated with the redistribution and reduction of consumers' incomes and the fluctuation of water authorities' revenues. Chapter X.3 introduces two environmental effects that have been noted. The two succeeding sections present the evidence relating to the possibilities discussed in X.2 and X.3, evidence which is derived from,

i)   The introduction over 1975 to 1981 by water authorities in England & Wales of the detailed trade effluent charging schemes for industry described in IV.2, above;

ii) The introduction in 1978 by the Perth Metropolitan Water Authority of a 'partial' user-pays pricing system, in which the 'standard charge' confers an allowance of 150 kilolitres per household per annum (approx. 125 lhd) before volumetric charges begin;

iii) The operation from 1981 onwards of the Federal Republic of Germany's 1976 Effluent Charge Law, which provided for the Lander to levy charges on direct dischargers for specified effluents into public waters; and

iv) The introduction in July 1982 by the Hunter District Water Board (Australia) of a full user-pays charging system for households (in an area with 400,000 population).

2. ECONOMIC EFFECTS: PRINCIPLES

Fundamental changes in pricing schemes for water services may have significant and very different effects on a) different groups of consumers and b) different individual consumers within a group or class.

Some may gain and some may lose; or, if a water service has hitherto been subsidised (e.g., the provision of irrigation water), all consumers may be losers to varying degrees while the general body of taxpayers reap the gains.

Important issues are raised by such changes. As discussed in Chapter II, any new charging system must make sense to consumers (i.e., both the reasons for it and the modus operandi must be at least broadly understood); it must also be seen by consumers to be equitable.

The distinction between different customer groups may be important. Thus if domestic consumers have been informed of the prospect of increasing expenditure on higher and higher cost supply sources which results from the absence of any incentive to economise on water use, they are more likely to accept the argument for metering or moves towards user-pays pricing. In any case, in developed countries, water charges seldom account for more than 1 per cent of households' disposable incomes. Although increased bills may result immediately from obligatory metering, the deferment or abandonment of expensive additional supply schemes will not be lost on householders. In commercial establishments, too, as long as equity among consumers is seen to be upheld, short-term increases in charges are likely to be accepted (the increase is less worrying if competitors are similarly affected).

In the industrial and agricultural sectors, however, reactions may be very different. Here the viability of an enterprise could be put at risk by the sudden and unexpected introduction of a 'rational' charging scheme for, for example, the public water supply, the discharge of trade effluents or the provision of irrigation water. Particularly vulnerable will be economic units

a) Which have water-intensive production methods (e.g., many farmers), and

b) Which are, because of the nature of their business (especially export or import-substitution), essentially price-takers and therefore unable to pass on increases in water charges.

In irrigation the problem may be particularly serious. The consequence of relatively low below-cost water charges for many years is that the value of this cheap resource will have been capitalised into land values. A sudden and permanent increase in water charges would mean large capital losses, either actual or potential, for those who had paid a high price for their land.

There are then three possible ways forward:

i) The water utility or authority may simply ignore the effects, and some enterprises may go out of business;

ii) The changes could be phased in slowly, perhaps over a period of many years;

iii) A separate fund, to relieve those households and firms and farmers hit particularly hard, could be established, by water authority or by government, to assit the 'victims' on a discretionary or formula basis.

Next we note the effects on a water authority's revenue in moving from a rate-based or tax-based charging system (financial risk low) to one based largely or wholly on volumetric charges (financial risk high). The problem for a utility arises if water sales fluctuate widely from year to year, most obviously due to the climate, with the consequence of revenue fluctuations. In order to cushion such fluctuations, the utility would have to create a special new reserve fund.

Finally, it should be realised that new ideas and new systems often seem to be very complex undertakings, requiring the 'education' of customers and staff alike. For example, a pricing scheme for effluent discharges, introduced through national legislation, would typically demand that regional or local water authorities acquire the technical capability to administer and oversee the operation of the scheme. The transitional administrative costs might be expected to be large.

3. ENVIRONMENTAL EFFECTS: PRINCIPLES

In some situations water service customers will be induced to substitute one (often inferior) source of supply for another following an alteration in relative prices. Thus if the price of the public water supply increases, consumers may be in a position to substitute their own direct abstractions (with pumping often the only cost, such is the nature of ground water rights). This may lead to originally unforeseen and longer-term effects on ground water levels and therefore on other parts of the supply system.

A second group of environmental effects noted by DeZellar and Maier (1980) are the possibly serious operating problems which may result from a sudden reduction in the volume of sewage effluent deriving from an economy-inducing water supply charging innovation. Two problems are said to occur because of reduced flows: "sedimentation resulting in accumulation of solids in sewer lines and anaerobic decomposition resulting in evolution of methane and hydrogen sulphide gases, with the latter causing corrosion and odour problems." (DeZellar and Maier (1980), p.79).

The more fundamental and underlying environmental benefits arise from the potential improvements in water resource management and related natural resources such as soil and forest. These of course overlap with the economic benefits of more efficient resource allocation.

4. ECONOMIC EFFECTS: EVIDENCE

How have these different types of effects worked out in practice? The Hunter Board in Australia made provision for both income redistribution and income fluctuation problems. Up to 1982-83, before the tariff changes, there was both inequity within the domestic sector and cross-subsidisation between sectors:

i) In 1981-82 5,594 low water consumers in low rateable valued areas paid an effective price of 99c. per kilolitre of water, while

ii) 464 high water consumers in similar low valued properties paid an effective price of 30c. per kilolitre,

iii) While the industrial sector, using about 36 per cent of water produced, only paid 23 per cent of total charges, the commercial sector used 50 per cent of total water and paid 50 per cent of charges.

Because to rectify the cross-subsidisation of industry would have a large impact on a small number of large industrial users, the Board decided that the user-pays system would be introduced into the non-residential sector in stages over a period three to five years ahead (Broad, 1984). To cope with the fact that a small minority of households (high water consumers, low valued property) would be particularly hard hit by the tariff change, especially in the first year while consumption patterns were still adjusting, a "Hardship Fund" was created and administered by trustees:

"It will provide assistance with meeting water and sewerage payments by households, and possibly by firms, whose positions are financially marginal. An amount of $A2m. should be provided in the first year." (Hunter District Water Board, 1982, p. 25). [Total annual revenue received by the Board in 1982-83 was expected to be $A68m.]

Secondly, the Hunter Board created a Revenue Fluctuation Reserve. The Board considered that under the new system the volume of water sold might fluctuate around its trend by about 10 per cent. Since, initially, the fixed charge in the user-pays structure was expected to generate 70 per cent of total revenue, a 10 per cent variation in variable charge revenue would result only in a 3 per cent fluctuation in total revenue. But this source of volatility would increase,

"as the balance between fixed and variable charges is changed, first by introduction of user-pays principles to non-residential properties, and then by a gradual move to a [fixed: variable charge] 50:50 split." (p.25).

Additionally, wet and dry years could come in runs, so a fund of more than 3 per cent of annual revenue would be required. It would be built up gradually.

The Perth authority (Western Australia) had the opposite problem to that of the Hunter Board: a significant non-domestic class subsidising the domestic sector. Introduction of full pay-for-use pricing for business/commercial customers, with parity (see II.2, above), would therefore have meant even more significant and rapid price increases for households than were already planned. This was not considered feasible, and therefore a gradual shift in the overall revenue burden towards the residential sector was deemed to be a necessary preliminary to full application of pay-for-service/pay-for-use pricing (see III.5, above). At the same time, the water authority in Perth was reported to have experienced financial difficulties because of the unexpectedly large drop in domestic consumption of 20 per cent to 30 per cent following tariff revision, with revenues falling dramatically.

In the German example, there are several provisions in the Waste Water Charges Act for the avoidance of unreasonable financial burdens. First, the charges have been phased in over a five-year period (see VII.3, above). Second, if additional waste water treatment plants are being planned or built by direct polluters (enterprises or communities) the charge may be greatly reduced in advance of commissioning for a period of three years. Finally, the Act included a hardship clause that permitted temporary exemptions if introduction of the charge, "would result in significant, detrimental economic consequences." (Brown and Johnson, 1984). An exemption could be for the whole charge or part of it. Brown and Johnson reported that, up to May 1984, eleven industries, several counties and a number of cities had petitioned the Minister of the Interior for exemptions. No results were yet known. According to work quoted by Brown and Johnson, earlier on -- in 1976 -- it had been envisaged that only in the pulp sector would the charge component loom large. And if some pulp factories were to be made financially unviable, "it would be seen as a modest advancement of the anticipated date of demise of old, technologically dated plants."

Similarly in the United Kingdom in 1979 a large amount of publicity was given to the much increased trade effluent charges faced by wool scouring firms in the Yorkshire Water Authority region. Here, again, the industry was already ailing.

Concerning irrigation, one report from Australia in 1984 described the effects of a 22 per cent rise in prices for irrigation water: "... whole townships shut down....marching on the state capital....international competitiveness with U.S. and Canada threatened."

On administrative costs, Brown and Johnson claim that "the difficulties of implementing [at the local level] a charges system were greatly overestimated.... After a few years of experience, several experts with substantial responsibility for administering the Effluent Charge Law [in Schleswig-Holstein, a rural area] have found it to be a far easier task than they had imagined, to their great surprise."

5. ENVIRONMENTAL EFFECTS: EVIDENCE

In Perth, where no controls existed on 'private' ground water abstraction when a partial user-pays system was introduced, there was a substantial switch from the public water supply to ground water. Because much of Perth's sewage is treated only by septic tanks, the potential for health risks (through contaminated supplies) increased, but not by much since most of the increased use of ground water was thought to be for garden watering.

A survey of the experience of 14 Californian wastewater treatment plants during the 1975-77 drought confirmed the second group of environmental effects noted in X.3. DeZellar and Maier (1980) concluded that,

"If decreased water use does become a normal way of life, efforts must be made to prepare for the attendant reduction in hydraulic loadings of wastewater collection and treatment systems. It is obvious from the limited sample of municipal sewer systems and treatment plants discussed in this paper that the potential impact of household water conservation on wastewater disposal systems is great. Sewer systems are subject to increased sedimentation, hydrogen sulphide generation, and clogging."

6. CONCLUSIONS

What lessons may be learned from these experiences?

First, great care must be taken in the implementation of new tariff structures. Radical changes may need to be phased in gradually, and attention must be paid to explaining the purpose and details of the new scheme to customers. Special provision may be required for these hardest hit.

Second, whereas householders and enterprises not engaged in international trade (e.g., commercial establishments, schools, hospitals, etc.) will be able to adjust over time to a new national charging scheme, some 'trading' producers will be in a much more difficult position. If, as in the examples mentioned, the result is an already contracting industry that simply contracts sooner, then compensation may be appropriate. If, however, a previously viable and 'stable' industry threatens to be seriously affected, then the trade-off is clear: an increase in efficiency in water use and in pollution loadings can be bought at the expense of industrial decline and higher unemployment in the short-term and, at best, industrial restructuring in the longer-term. In some Member states that price may be willingly paid; in others, it may be felt that concerted international action on pricing structures for the main water services to industry is a higher priority.

Chapter XI

CONCLUSIONS

In Chapter II of this report it was stated that ideally water services should be provided in an economically and environmentally efficient manner so that the net benefits to the community as a whole from their provision are maximised. The case for marginal cost pricing follows logically from this objective. The 'marginal cost' referred to should include <u>all</u> relevant costs: natural resource depletion costs (the opportunity costs of using the basic resource itself), damage costs, and the more familiar resource costs of operating, maintaining and replacing when necessary the supply system.

It was recognised, however, that attention must also be given, in the design of tariff structures, to a number of other important considerations. Foremost among these are the pursuit of fairness (both among different classes of consumers and, within classes, among individual consumers) and the obligation to raise sufficient overall revenue to satisfy what are sometimes externally-imposed financial requirements. Additionally, there is the need to command consumer acceptance and confidence, the avoidance of any significant danger to public health, and the minimisation of administrative costs. This last obligation imposes limits on the feasible sophistication of charging systems designed to maximise allocative efficiency (in the economic and environmental sense) but does not rule out the construction of a wide variety of tariff structures based on efficiency criteria, as the experience of the period since 1970 has shown.

In succeeding sections of the report (Chapters III to VIII) the major water services -- piped supply and disposal, and direct abstractions and discharges -- were explored in much more detail and the relevant experiences of OECD Member states were assembled and compared. Two phenomena were highlighted by this evidence. First, the inefficiency of flat-rate pricing systems, when <u>no</u> volumetric charge confronts the consumer (<u>switching</u> to a metering system was dealt with formally in Chapter VIII). Second, the prevalence of various forms of average cost pricing was noted, often incorporated into a two or multi-part tariff. In a growing number of countries increasing-block tariffs are gaining in popularity, and the results claimed in terms of a levelling-off or reduction in consumption are often impressive.

Although not satisfying fully efficiency requirements, the political attractiveness of increasing-block tariffs was recognised. Indeed, in an increasing-cost industry the highest-priced block may be fixed approximately at marginal cost (as has happened in Japan) and lower-priced tranches used to ensure that gross revenue does not exceed financial requirements. Whatever the objectives and determinants of volumetric charging, however, the general superiority of this type of user-pays tariff over a flat-rate charging system is clear. For the User-Pays Principle, which embraces the more familiar Polluter-Pays Principle, ensures that a financial incentive exists for the water service user to avoid waste. This remains true even in the presence of sometimes still significant subsidies, especially of capital costs (as noted in Chapter IX). But a general decline of subsidisation has also been in evidence.

A change towards a more rational tariff structure is often not a painless process. Chapter X of the report listed the income redistributional and environmental costs that may result and the methods by which water authorities and water legislation have attempted to cope with such problems.

Annex 1

WATER RESOURCES AND OPPORTUNITY COSTS

Because most natural resources do not become available how and where consumers would prefer to make use of them, capital and human resources are put to work, often on a very large scale indeed, to overcome such imbalances of form, quality and location. Water has an added imbalance, of a temporal nature: it is most often desired when precipitation and natural flows are at their lowest. As well as collection, transportation, purification and distribution, then, water must also be subject to storage. In this way -- by the application of complementary but scarce resources -- water is transformed into an economic good. A similar sequence of events gives piped sewage disposal and treatment its economic dimension.

By identifying, measuring and valuing in economic terms the capital, human and other resources committed, it is possible to evaluate the opportunity costs of providing a reliable water supply or sewage disposal service. These form an important focus when the basis of charging policies is considered.

So far the concern has been, by implication, only with fabricated capacity to supply or dispose of water; thus only piped water supplies and piped effluent disposal (and treatment) have been considered. What of the natural capacity of waters (rivers, lakes, aquifers, etc.) to supply water and to assimilate effluents? Does the use of such 'services' not constitute a free gift of nature, lacking any opportunity costs and therefore suggesting free distribution, i.e., a zero charge?

In order to answer this question, suppose first that there exists a natural body or flow of water

a) Which is unaugmented, directly or indirectly, by any fabricated works,

b) The claims on which, for supply or disposal purposes, do not in aggregate exceed reliable availability, and

c) The use of which has no effect on the enjoyment or production possibilities of other users, actual or potential, i.e., there are no externalities.

In such an 'ideal' situation, quality would be maintained and the demands of any one time period (day, year, generation, etc.) would have no consequences for succeeding time periods. There would thus be no damage costs and no quantity or quality depletion costs. If any of (a) to (c) are relaxed, however, the opportunity cost of use becomes a relevant and indeed indispensable concept. If (i) reliable availability can only be maintained with the aid of capital works (barrages, reservoirs, etc.), or (ii) the claims in aggregate do put a strain on availability (and thus make the supply less reliable), or (iii) qualitative or quantitative deterioration affects other economic agents (consumers or producers), then opportunity costs will be incurred as a result of use: resource-use costs, depletion costs or damage costs, respectively.

In the most general cases, therefore, the opportunity costs of the provision of a water supply or effluent disposal service, natural or fabricated, may be defined as:

Opportunity Costs = Resource-use costs (i.e., the goods or services forgone by the commitment of economic resources in the construction or operation of the supply system)
+ Natural resource depletion costs (qualitative or quantitative)
+ Damage costs.

Economic progress in general and technological advances in particular have meant that in the developed world the situation described by (a), (b) and (c), above, is now extremely rare. Indeed, it may well be extinct. The demands on natural waters have largely outstripped the natural capacity to provide (water) or to dilute (effluents). Growing demands have meant that either quantity is reduced or, if effluents are discharged, water quality is downgraded. Enormous damage to a river system may occur as large quantities of wastes build up. Part or complete restoration to the original state, or even just maintenance of the system at some given imperfect state, will require the commitment of economic resources, often on a significant scale.

Annex 2

THE CALCULATION OF MARGINAL COST

The methodology adopted in the text (section III.6) for calculating marginal cost in a 'systems' industry is rooted in the measurement of the effect upon system costs of faster or slower growth in the quantity of the commodity to be supplied.

Thus in Turvey's original article (1969) the marginal cost for any year

"... is the excess of (a) the present worth in that year of system costs with a unit permanent output increment starting then, over (b) the present worth in that year of system costs with the unit permanent output increment postponed to the following year." (page 289).

The unit output increment must be "large enough to be noticeable but small enough to be marginal" (p.288). Clearly it will contain a capital, or capacity, element "only in those years when demand is at or near the reliable yield of existing capacity" (Turvey, 1976, p.161). Indeed, "[I]n between successive schemes, when reliable yield much exceeds demand, capacity costs are zero" (ibid).

The problem with this pursuit of 'near-horizon' allocative efficiency signals (i.e., looking ahead one year at a time) has been stated in section II.1 (in main text). Because some of todays' investment choices by water service consumers (e.g., to install recirculation equipment or not, to order this cooling system or that one, perhaps with quite different water-use characteristics) will have implications for many years to come, the present signal concerning the price of water should embody information about the cost of water over a number of years ahead. For practical purposes therefore, the marginal capacity cost (MCC) should be calculated as an 'average' of the MCCs derived as above for a number of future years. The average would be calculated (with the aid of a suitable discount rate) as the annual equivalent of the stream of derived MCCs.

The best method of calculating such an average, if the appropriate cost and demand data are available, is by defining the marginal cost in this way:

131

Average Marginal Incremental Cost (AMIC)
    = Marginal Operating Cost (MOC) + Marginal Capacity Cost (MCC),

(1)   i.e. AMIC = MOC +

$$\text{AMIC} = \text{MOC} + \frac{\left(\begin{array}{l}\text{Present Worth (PW) of} \\ \text{System Costs with one} \\ \text{planned expansion}\end{array}\right) - \left(\begin{array}{l}\text{PW of System Costs} \\ \text{with a different} \\ \text{planned expansion}\end{array}\right)}{(\text{PW of difference in quantities of water delivered})}$$

The two PWs in the numerator on the right-hand side might arise as a result of detailed system investment appraisal which has included sensitivity testing of the chosen investment programme with respect to changes in demand. Again, the difference in the planned expansions of the supply system (i.e., the difference in the demand forecast profiles) should ideally be large enough to be noticeable but small enough to be marginal. One demand forecast should be the central estimate of the growth in water demand, and the other a slightly (i.e., marginally) higher or lower demand projection.

If, however, no sensitivity testing has been or can be undertaken, it will generally be necesssary to calculate marginal cost as

Average Incremental Costs (AIC) =   Marginal Operating Cost (MOC) +
                                   + Marginal Capacity Cost (MCC),

(2)   i.e. AIC = MOC +

$$\text{AIC} = \text{MOC} + \frac{\begin{array}{l}\text{PW of the Least Cost Stream of Investment} \\ \text{Expenditures (those sensitive to quantity of water} \\ \text{use)}\end{array}}{\begin{array}{l}\text{PW of the Stream of Incremental Output resulting} \\ \text{from the Investment Stream}\end{array}}$$

This information on the costs and quantities of water associated with the least cost investment programmed will similarly usually have been generated in investment appraisal work. MCC calculation techniques which defer increases in water demand and the associated investment programme by one year (e.g., Department of Environment/National Water Council, 1980) give estimates which are identical to those produced by equation (2).

In the absence of information about expansion costs of the whole system resort must be made to the textbook definition of marginal cost. An investment price signal would be obtained by estimating MCC as the equivalent annual costs per unit of output of the next 'lump' of investment. Thus

(3)   Textbook Long-Run Incremental Cost (TLRIC) = MOC + $\dfrac{r.I_k}{Q_{k+1} - Q_k}$

where I_k is the cost of the next major lump of investment (in year k), r is the capital recovery factor (equal to the annuity that will repay a £1 loan over the useful life of the investment with compound interest equal to the opportunity cost of capital on the unpaid balance), and Q_k, Q_{k+1}, are the volumes of water produced in years k and k+1 (World Bank, 1977, pp. 23-4).

Finally, it should be noted that it is possible to apply the AMIC or AIC methodology in order to estimate the marginal capacity cost of any chosen peak. Suppose the MCC of peak-hour capacity was required. Then

(4) AMIC (of peak-hour capacity)

$$= \frac{\left(\begin{array}{l}\text{PW of System Costs with} \\ \text{planned expansion, and peak-hour} \\ \text{capacity higher than expected}\end{array}\right) - \left(\begin{array}{l}\text{PW of System Costs} \\ \text{with planned and} \\ \text{peak-hour capacity} \\ \text{as expected}\end{array}\right)}{(\text{PW of future years' increments in peak-hour capacity})}$$

= the 'average' cost of the marginal increase in peak-hour capacity, expressed on an annual basis.

133

Annex 3

ECONOMIC THEORY OF THE METERING DECISION

The economic theory of the metering decision is presented here in the context of the domestic use of water, although the same analysis may also be applied to other sectors.

A. ECONOMIC APPRAISAL

IMPACT EFFECT OF METERING

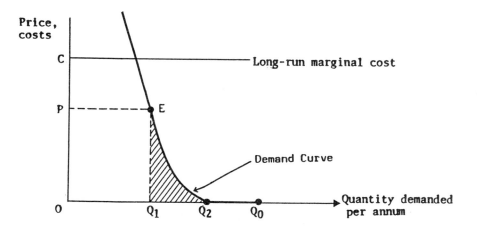

The diagram illustrates the situation before and after the introduction of domestic meters and a volumetric charging system. For the moment we abstract from the possibility of peaks and seasonal demand variation, i.e., we assume that demands are spread uniformly over time, both before and after metering.

We assume that the demand curve is for an 'average' house-hold. Before metering, price is zero and demand is running at the rate Q_0. After metering it is assumed that price rises to P and demand falls to Q_1. $Q_2 Q_0$ is 'useless' consumption which is assumed to be eliminated by the household when any positive price is charged. Assume the long-run marginal cost of supplying and disposing of water to be constant at C. The question is then simple: do the benefits of a changeover to metering and volumetric charging exceed the costs?

There are two major benefits if we ignore the intangible advantage that some would stress as the most important gain: the greater fairness that arguably results from charging by volume.

     i) <u>Reduction in water service costs</u>. A reduction in household water use, from Q_0 to Q_1, will give rise to

    a) A decrease in the operation and maintenance costs of existing water supply and sewage disposal facilities, and, much more important,

    b) The postponement of new works in water supply, sewerage and sewage disposal.

The present worth of this benefit is the excess of the present worth of the current and capital costs stream ensuing without metering over the present worth of the corresponding cost stream associated with metering.

Alternatively $S = C.R$ where S = annual benefits per household
R = annual reduction in water demands by the household
and C = marginal cost of supply and disposal (in the diagram $C = OC$ and $R = Q_1 Q_0$).

Note that C is made up of

    a) Water conservation and bulk delivery,
    b) Reinforcement of the distribution system,
    c) Reinforcement of the sewerage system, and
    d) Trunk sewers and sewage treatment.

    ii) <u>Better forecasting and waste reduction</u>. Metering of all or part of a distribution system provides additional useful information for the water authority; in particular it should be able to improve its demand forecasting and waste reduction techniques and performance. This is difficult to value. Call it I.

There are also two major costs of metering:

    iii) <u>Resource costs of metering</u>. These will include

    a) Provision of installation (differing for internal and external installation)),
    b) Maintenance, and
    c) Additional reading, billing and collection costs over and above those incurred with the existing charging system.

Let (a), (b) and (c) sum to the equivalent of M per annum.

iv) <u>Value of consumption of water forgone</u>. Let this be U per annum. U may be measured as the area under the demand curve between Q1 and Qo: the shaded area in the diagram.

(Note that we do not include as a separate item the costs associated with actually reducing the use of water, i.e., the inconvenience and repairs expenditure that members of the household suffer in their attempts to reduce consumption. These are already reflected in the demand curve, such 'involuntary' demands reflecting the fact that, for example, a consumer is assumed to be willing to pay a certain small amount -- say 2 pence per 1 000 litres -- for water being lost through a leaking tap if the cost to him of repairing that leak is more than 2 pence 1 000 litres.)

Formally, the metering decision-rule can be set out with either all the benefits and costs in present worth terms or in annual equivalent terms. Using the latter, on economic efficiency grounds metering and volumetric charging is to be recommended if

$$C . R + I > M + U$$

In a situation of increasing water demands over time, the exercise is probably best undertaken largely in present worth terms. It is then still possible to cast the inequality in annual equivalent terms, once a discount rate has been specified.

We now recognise the reality of seasonal and peak demands, and the consequent desirability of distinguishing between the costs of water supply and the costs of sewerage and sewage disposal (S&SD). The various peaks and the demands of particular seasons may be affected very differently by metering, depending on the structure of demands associated with each particular peak or season. For example, peak-hour domestic use, usually caused by domestic irrigation during very hot and dry weather, would be expected to be cut back much more than would winter or even 'year round' demand (for evidence, see Tables 15). The benefits of reductions in peak-hour public water supply consumption are the product of the quantitative saving and the marginal cost of peak-hour provision, which will measure the effect on system costs of expansions or contractions in peak-hour demands. Annex 2 sets out a methodology for estimation.

Acceptance of this non-uniformity of demand means that the "C.R" factor in the inequality above needs to be replaced by the summation of a number of different factors, referring to the various seasonal and peak demands affected and their associated marginal costs. Water supply and S&ST should be treated separately since, for example, reduction of garden irrigation demands leads to no saving in S&SD costs. Thus the metering decision rule becomes: domestic metering is to be recommended on economic efficiency grounds if

$$\Sigma C_i - R_i + I > M + U$$

where
C_i = long-run marginal cost of water supply (and S&SD) provision in peak or off-peak period (or season) i,

R_i = reduction in water supply (or S&SD) demands in peak (or off-peak) period (or season) i,

I = benefits of improved waste reduction and demand forecasting (each year),

M = direct resource costs of metering and volumetric charging (each year),

U = the value of water consumption forgone (each year), and

n = the number of seasons or periods into which the year is divided.

B. FINANCIAL APPRAISAL

A water utility will also be concerned with the effects of domestic metering on cash inflow and outflow, i.e., with financial appraisal. In general, the financial cost to the utility of resources and factors of production and the financial benefits generated by operating cost savings and capital deferment may be equated with the economic costs and benefits already discussed in (A). The main differences that emerge in a financial appraisal are

i)   Instead of the opportunity cost of public sector capital, the discount rate used should be the utility's inflation-adjusted borrowing rate of interest,

ii) The 'loss' of water to consumers (the 'U' factor) would not appear, since no compensation would be paid by the utility, and

iii) Because charges would, after metering, be paid in arrears to the utility instead of in advance, the utility would lose interest receipts on its (lower) cash balance (note, however, that consumers would be correspondingly better off because of later cash payments).

C. EQUITY

One distributional implication should be noted. With metering, the incidence of charges among consumers will be different. For example, poor, large families would probably pay more. If the government (or even the water authority) has particular income distribution objectives it may wish to take account of such effects. Governments may counter undesirable distributional effects through tax and social security policies and/or by instructing authorities to offer lifeline rates, to alter the incidence of standing charges, etc.

NOTES AND REFERENCES

1. Consumer's surplus represents the excess of what a consumer would be prepared to pay for the service and the amount of water consumed over what in fact is paid. If the utility tries to recover more than consumer's surplus, then it will cause the consumer to disconnect from the system, which will be inefficient (i.e., economically unjustified) if the utility is already recovering through fixed charges the cost of keeping the consumer in the system.

2. Financial risk may be formally defined as the proportion of overall expected revenue depending on the actual quantity of water consumed.

3. We define a systems industry as one in which the various 'production plants' (or sources of supply) are integrated to such a degree that it is more sensible to speak of the system producing output rather than, for example, a particular reservoir.

4. Note, however, that if the fixed element of the tariff is consumer-related e.g., being geared to the width of the service pipe in the provision of the public water supply, then equiproportionate reductions in all the fixed elements would tend to compress, in the extreme case to zero, the cost differentials associated with different pipe sizes. This would in theory lead to economically inefficient selection of pipe sizes by consumers, although in practice the extent of the consequent resource misallocation would probably be small.

5. Separate metering or, more inexpensively, estimation of garden water supplies (not disposed of into the sewerage system) would permit the supplement to be more closely linked to the volume of sewage generated. If users augment the public water supply with their own abstractions, water authorities may need to levy extra S&SD charges and/or require additional metering.

6. For sewers in United Kingdom urban areas at 3 metres depth an increase in volumetric capacity of 300 per cent is associated with an increase in annual equivalent capital costs of only 65 to 90 per cent.

7. It is unclear whether optimising models for the design of sewage works which allow such analysis and estimation are yet available for general application.

8. We are neglecting the unlikely prospect of excess supplies stretching for a long time into the future; for this situation a near-zero

authorised abstraction charge would be in order.

9. It is of course recognised that in practice damage functions may be too costly to estimate and that marginal costs can often at best be inferred from complaints and observations.

10. American Water Works Association, 1957. Water Rates Manual (New York, A.W.W.A.).

11. Anderson, T.L. (ed), 1983. Water Rights (Cambridge, Mass., Ballinger).

12. Barden, D.S., & Stepp, R. J., 1984. "Computing Water System Development Charges", Journal American Water Works Association, Vol. 76, No. 9.

13. Benson, V.W., Landgren, N.E. & Cotner, M.L., 1979. Irrigation Water Management in an Energy Short Economy (paper transmitted to the U.N. Economic Commission for Europe Seminar on Rational Utilisation of Water at Leipzig, September 1979, mimeo).

14. Boland, J.J., 1983. Water/Wastewater Pricing and Financial Practices in the United States (Washington, D.C., MetaMetrics Inc., mimeo).

15. Bradshaw, J., & Harris, T. (eds.), 1983. Energy and Social Policy (London, Routledge Kegan Paul).

16. Broad, P., 1984. Water Pricing: The Hunter Experience (Hunter District Water Board, mimeo).

17. Brown, G.M., & Johnson, R.W., 1984. "Pollution Control by Effluent Charges: It Works in the Federal Republic of Germany, Why Not in the U.S.", Natural Resources Journal, Vol. 24.

18. Coe, A.L., 1978. Water Supply and Plumbing Practices in Continental Europe (London, Hutchinson Benham).

19. Davis, R.K. & Hanke, S.H., 1971. Planning and Management of Water Resources in Metropolitan Environments (Washington, George Washington University Natural Resources Policy Centre).

20. Department of the Environment, 1974. The Water Services: Economics of Financial Policies. Third Report to the Secretary of State for the Environment (London, Her Majesty's Stationery Office).

21. Department of the Environment, 1985. Joint Study of water Metering: Report of the Steering Group (London, Her Majesty's Stationery Office).

22. Department of the Environment/National Water Council, 1980. Leakage Control Policy and Practice (London, National Water Council).

23. Department of Resources and Energy, 1983. Water 2000: Consultants Report No. 3 Economic and Financial Issues (Canberra, Australian Government Publishing Service).

24.   DeZellar, J.T., & Maier, W.J., 1980. "Effects of Water Conservation on Sanitary Sewers and Wastewater Treatment Plants", Journal of the Water Pollution Control Federation, Vol. 52, Part 1.

25.   Dirickx, J. & Tritsmans-Sprengers, M., 1985. Experiences with Water Rates at the Antwerp Water Works, paper delivered at the 1985 A.S.P.A.C. Seoul Conference (mimeo).

26.   Elliot, R.D., 1973. "Economic Study of the Effect of Municipal Sewer Surcharges on Industrial Wastes and Water Usage", Water Resources Research, Vol. 9, No. 5.

27.   Elthridge, D.E., 1970. "An Economic Study of the Effect of Municipal Sewer Surcharges on Industrial Wastes" (Raleigh, North Carolina State University, unpublished Ph.D. thesis).

28.   Environmental Defense Fund, 1980. Water for Denver: An Analysis of the Alternatives (Colorado, Environmental Defense Fund).

29.   Federal Ministry of the Interior, 1983. Report on the Experience with the Waste Water Charges (Bonn, Federal Ministry of the Interior).

30.   Feldman, S.L., 1975. "Peak-Load Pricing through Demand Metering", Journal of the American Water Works Association, Vol. 67, No. 9.

31.   Finnish Energy Economy Association, 1983. Report ETY 9/1983 Experiences of charging of hot water consumption in renovated buildings (Helsinki, Finnish Energy Economy Association, mimeo).

32.   Fleer, H.-D., 1979. The Use of Domestic Wastewater for Irrigation (paper transmitted to the U.N. Economic Commission for Europe Seminar on Rational Utilisation of Water, at Leipzig, September 1979, mimeo).

33.   Flinn, J.C., 1969. "The Demand for Irrigation Water in an Intensive Irrigation Area", Australian Journal of Agricultural Economics, Vol. 13, pt. 3.

34.   Frankham, J. & Webb, M.G., 1977. "The Principle of Equalisation and the Charging for Water", Public Finance and Accountancy, Vol. 4, No. 6.

35.   Gallagher, D.R., Boland, J.J., le Plastrier, B.J. & Howell, D.T., 1981. Methods for Forecasting Urban Water Demands, Australian Water Resources Council Technical Paper No. 59 (Canberra, Australian Government Publishing Service).

36.   Gallagher, D.R. and Robinson, R.W., 1977. Influence of Metering, Pricing Policies and Incentives on Water Use Efficiency, Australian Water Resources Council Technical Paper No. 19 (Canberra, Australian Government Publishing Service).

37.   Gilling, W.J.W., 1980. "La Politique de la Banque Mondiale en Matière d'eau", Aqua, No. 2.

38.   Griffith, F.P., 1982. "Policing demand through pricing", Journal American Water Works Association, Vol. 74, Part 6.

39. Grima, A.P., 1972. Residential Water Demand: Alternative Choices for Management (Toronto, University of Toronto Press).

40. Gysi, M., 1981. "The Cost of Peak Capacity Water", Water Resources Bulletin, Vol. 17, No. 6.

41. Gysi, M., & Lamb, G., 1977. "An Example of Excess Urban Water Consumption", Canadian Journal of Civil Engineering, Vol. 4, pt. 1

42. Hanemann, M., 1978. Report to E.P.A. on a Study of Industrial Responses to Water and Waste Charges (Berkeley, University of California Press).

43. Hanke, S.H., 1972. "Pricing Urban Water" in Mushkin, S.J. (ed.), Public Prices for Public Products (Washington, D.C., Urban Institute).

44. Hanke, S.H., 1975. "Water Rates: An Assessment of Current Issues", Journal of the American Water Works Association, Vol. 67, No. 5.

45. Hanke, S.H., 1978. "A Method for Integrating Engineering and Economic Planning", Journal American Water Works Association, Vol. 70, No. 9.

46. Hanke, S.H., 1980. "L'économie réelle de la lutte contre le gaspillage de l'eau", Techniques et Sciences Municipales, Vol. 75, No. 2.

47. Hanke, S.H., 1981. "On the Marginal Cost of Water Supply", Water Engineering and Management, February.

48. Hanke, S.H. & de Maré, L., 1982. "Residential Water Demand: A Pooled, Time Series, Cross Section Study of Malmö, Sweden", Water Resources Bulletin, Vol. 18, No. 4.

49. Hanke, S.H. & Wenders, J.T., 1982. "Costing and Pricing for Old and New Customers", Public Utilities Fortnightly, 29 April 1982.

50. Hanke, S.H. & Wentworth, R.W., 1981. "On the Marginal Cost of Wastewater Services", Land Economics, Vol. 57, No. 4.

51. Herrington, P.R., 1980. Mantaro Transfer Project: Submission to Binnie & Partners, Consulting Engineers. Note 3: Calculation of Marginal System Opportunity Costs of Bulk Water Supplies (University of Leicester, mimeo).

52. Herrington, P.R., 1982. "Water: A Consideration of Conservation", Journal of the Royal Society of Arts, Vol. cxxx, No. 5310.

53. Herrington, P.R., & Tate, J.C., 1971. The Metering of Residential Water Supplies: Some Empirical Evidence, Water Demand Study Progress Paper S.2 (revised) (Leicester, University of Leicester, Department of Economics).

54. Herrington, P.R., & Webb, M.G., 1981. "Charging Policies for Water Services", Water Services, Vol. 85, No. 1025.

55. Hirshleifer, J., DeHaven, J.C. & Milliman, J.W., 1960. Water Supply: Economics, Technology, and Policy (Chicago, University of Chicago Press).

56. Howe, C.W., 1982. "The Impact of Price on Residential Water Demand: Some New Insights", Water Resources Research, Vol. 18, No. 4.

57. Howe, C.W., & Linaweaver, F.P., 1967. "The Impact of Price on Residential Water Demand and its Relation to System Design and Price Structure", Water Resources Research, Vol. 3, No. 1.

58. Hunter District Water Board, 1982. An Equity-Based Water and Sewer User-Pays Tariff (Hunter District Water Board, mimeo).

59. International City Management Association, 1970. "Sewer Services and Charges", Urban Data Services, Vol. 2, No. 2.

60. Jean, M., 1980. L'application de la théorie du coût marginal au tarif de vente d'eau dans un ouvrage à buts multiples (paper transmitted to the UN Economic Commission for Europe Seminar on Economic Instruments for Rational Utlisation of Water Resources at Veldhoven, October 1980, mimeo).

61. Jenking, R.C., 1973. Fylde Metering (Blackpool, Fylde Water Board)

62. Keller, C.W., 1977. "Pricing of Water", Journal of the American Water Works Association, Vol. 69, No. 1.

63. Kinnersley, D., 1980. Water Use and Consumption (London, International Water Supply Association).

64. Kinnersley, D., 1982. The Economy of River Basins (London, mimeo).

65. Langord, J.J., & Heeps, D.P., 1985. A Demand Management Strategy for Victoria (mimeo).

66. Larkin, D.G., 1978. "The Economics of Water Conservation", Journal American Water Works Association, Vol. 70, No. 9.

67. Laukkanen, R., 1981. Flow Forecasts in General Planning of Municipal Water and Sewage Works, National Board of Waters Water Research Institute Publication No. 41 (Helsinki).

68. Lippiatt, B.C, & Weber, S.F., 1982. "Water Rates and Residential Water Conservation", Journal American Water Works Association, Vol. 74, No. 6.

69. Loudon, R.M., 1984. "Region of Durham Experiences in Pricing and Water Conservation", Canadian Water Resources Journal, Vol. 9, No. 4.

70. Mann, P.C., & Schlenger, D.L., 1982. "Marginal cost and seasonal pricing of water service", Journal American Water Works Association, Vol. 74, Part 1.

71. Martin, W.E., Ingram, H.M., Laney, N.K., & Griffin, A.H., 1984. Saving Water in a Desert City (Washington, D.C., Resources for the Future).

72.   Martini, P.L., 1980a. Water Use and Consumption, paper delivered at 1980 IWSA Paris Conference (London, International Water Supply Association).

73.   Martini, P.L., 1980b. Finance and Water Tariffs, paper delivered at 1980 IWSA Paris Conference (London, International Water Supply Association).

74.   McLamb, D.W., 1978. Industrial Demand for City-supplied Water and Waste Treatment (Raleigh, North Carolina State University, unpublished dissertation).

75.   Metropolitan Water Authority, 1985. Domestic Water Use in Perth, Western Australia (Leederville, Metropolitan Water Centre).

76.   Milliman, J.W., 1968. General Economic Principles for Sewerage Planning and Operation (Bloomington, Indiana University Graduate School of Business Institute for Applied Urban Economics).

77.   Mohle, Mr., 1979. Problems and Limits of Saving Drinking Water by Construction of Double Supply Grids and Watersaving Installations and Fittings at the Consumer's End (paper transmitted to the UN Economic Commission for Europe Seminar on Rational Utilisation of Water, at Leipzig, September 1979, mimeo).

78.   National Agency of Environmental Protection, 1984. Conclusions of the Water Price Committee (Copenhagen, Ministry of the Environment, mimeo).

79.   O'Neill, P.G., 1971. The Implementation of Charging Schemes by River Authorities in South-East England. Report by Mrs J.A. Rees to the Water Resources Board (London, Department of the Environment, mimeo).

80.   National Water Council, 1976. Paying for Water (London, National Water Council).

81.   Ng, Y-K., 1983. Quasi-Pareto Social Improvements (Monash University, Department of Economics, Seminar Paper 9/38, mimeo).

82.   OECD, 1976. Water Management in France (Paris, OECD Environment Directorate, mimeo).

83.   OECD, 1977. Water Management Policies and Instruments (Paris, OECD).

84.   OECD, 1980. Water Management in Industrialised River Basins (Paris, OECD).

85.   Rees, J.A., 1981. "Irrelevant Economics: The Water Pricing and Pollution Charging Debate", Geoforum, Vol. 12, No. 3.

86.   Rees, J.A., 1982. An Economic Approach to Waste Control: A Second Look, paper delivered at the IWES Symposium on An Understanding of Water Losses (London).

87.   Rees, R., 1984. Public Enterprise Economics, 2nd edition (London, Weidenfeld & Nicolson).

88. Rice, I.M. & Shaw, L.G., 1978. "Water Conservation - A Practical Approach", Journal of the American Water Works Association, Vol. 70, No. 9.

89. Rotterdam Water Authority, 1976. Interim Report of Working Party on Rotterdam Water Authority's Sales Forecast, in Dutch (Rotterdam).

90. Russell, J.D., 1984. "Seasonal and Time-of-Day Pricing", Journal American Water Works Association, Vol. 76, No. 9.

91. Sagoh, M., Aya, H. & Funaki, M., 1978. Reuse of Water and Recycling, paper delivered at the 12th IWSA Congress in Kyoto, (London, International Water Supply Association).

92. Sewell, W.R.D. & Roueche, L., 1974. "Peak Load Pricing and Urban Water Management", Natural Resources Journal, Vol. 13, pt. 3.

93. Sharpe, W.E., 1978. "Why Consider Water Conservation?", Journal American Water Works Association, Vol. 70, No. 9.

94. Shipman, H.R., 1978. Water Metering Practices, Aqua, No. 2.

95. Snijders, J.A.C., 1980. Contribution of the Netherlands to the General Report 'Water Use and Consumption' (Drinkwaterleiding Rotterdam, mimeo).

96. South West Water Authority, 1975. Water Conservation (Exeter, South West Water Authority, mimeo).

97. Speed, H.D.M., 1980. Energy Saving (London, International Water Supply Association.

98. Takesada, K., 1980. "Sewer User Charges in Osaka", Journal of the Water Pollution Control Federation, Vol. 52, No. 5.

99. Tate, D.M., 1980. Economic Incentives for the Rational Use of Water Including Protection Against Pollution (paper transmitted to the UN Economic Commission for Europe Seminar on Economic Instruments for Rational Utilisation of Water Resources, at Veldhoven, October 1980, mimeo).

100. Thackray, J.E., 1976. "Charging for the direct discharge of effluents to rivers and aquifers", Chemistry and Industry, No. 19.

101. Thackray, J.E., Cocker, V. & Archibald, G., 1978. "The Malvern and Mansfield Studies of Domestic Water usage: Discussion", Proceedings of the Institution of Civil Engineers, Vol. 64.

102. Thackray, J.E. & Archibald, G.G., 1981. "The Severn-Trent Studies of Industrial Water Use", Proceedings of the Institution of Civil Engineers, Part 1, Vol. 70.

103. Thames Water, 1978. Future Charging for Piped Supply and Disposal Services (Thames Water Authority, London, mimeo).

104. Thomas, J.F., Syme, G.J., & Gosselink, Y.F., 1983. Household Responses to Changes in the Price of Water in Perth, Western Australia, paper delivered at the Hobart Hydrology and Water Resources Symposium, November 1983 (mimeo).

105. Turvey, R., 1969. "Marginal Cost", Economic Journal, Vol. LXXIX.

106. Turvey, R., 1976. "Analysing the Marginal Cost of Water Supply", Land Economics, Vol. 52, No. 2.

107. United Nations, 1976. The Demand for Water: Procedures and Methodologies for Projecting Water Demands in the Context of Regional and National Planning. (New York, United Nations).

108. United Nations, 1980. Efficiency and Distributional Equity in the Use and Treatment of Water: Guidelines for Pricing and Regulations (New York, United Nations).

109. Vermersch, R., 1974. Influence du comptage sur la consommation d'eau et les moyens d'améliorer son efficacité (London, International Water Supply Association).

110. Victorian Water and Sewerage Authorities Association, Institute of Water Administration and Victoria Department of Water Resources, 1985. Reform of Urban Water Tariff Structures, incl. paper by C. G. Pollett, "Tariff and Revenue Policy Reform in the Water Industry in Western Australia" (appendix).

111. Water Resources Board, 1972. Desalination 1972 (London, Her Majesty's Stationery Office).

112. Water Resources Board, 1974. Water Resources in England and Wales. Volume 2: Appendices (London, Her Majesty's Stationery Office.

113. Webb, M.G. & Woodfield, R., 1981. "Standards and Charges in the Control of Trade Effluent Discharges to Public Sewers in England and Wales", Journal of Environmental Economics and Management, Vol. 8.

114. World Bank, Staff Working Paper No. 259, 1977. Alternative Concepts of Marginal Cost for Public Utility Pricing: Problems of Application in the Water Supply Sector (Washington, International Bank for Reconstruction and Development).

115. Young, J.A., 1976. Control of Water Supply Demand, paper delivered at the 11th IWSA Congress in Amsterdam, September 1976 (London, International Water Supply Association).

116. Zamora, J., Kneese, A.V. & Erikson, E., 1981. "Pricing Urban Water: Theory and Practice in Three Southwestern Cities", The Southwestern Review of Management and Economics, Vol. 1, No. 1.

OECD SALES AGENTS
DÉPOSITAIRES DES PUBLICATIONS DE L'OCDE

ARGENTINA - ARGENTINE
Carlos Hirsch S.R.L.,
Florida 165, 4° Piso,
(Galeria Guemes) 1333 Buenos Aires
Tel. 33.1787.2391 y 30.7122

AUSTRALIA-AUSTRALIE
D.A. Book (Aust.) Pty. Ltd.
11-13 Station Street (P.O. Box 163)
Mitcham, Vic. 3132   Tel. (03) 873 4411

AUSTRIA - AUTRICHE
OECD Publications and Information Centre,
4 Simrockstrasse,
5300 Bonn (Germany)   Tel. (0228) 21.60.45
Local Agent:
Gerold & Co., Graben 31, Wien 1 Tel. 52.22.35

BELGIUM - BELGIQUE
Jean de Lannoy, Service Publications OCDE,
avenue du Roi 202
B-1060 Bruxelles   Tel. (02) 538.51.69

CANADA
Renouf Publishing Company Ltd/
Éditions Renouf Ltée,
1294 Algoma Road, Ottawa, Ont. K1B 3W8
Tel: (613) 741-4333
Toll Free/Sans Frais:
Ontario, Quebec, Maritimes:
1-800-267-1805
Western Canada, Newfoundland:
1-800-267-1826
Stores/Magasins:
61 rue Sparks St., Ottawa, Ont. K1P 5A6
Tel: (613) 238-8985
211 rue Yonge St., Toronto, Ont. M5B 1M4
Tel: (416) 363-3171
Sales Office/Bureau des Ventes:
7575 Trans Canada Hwy, Suite 305,
St. Laurent, Quebec H4T 1V6
Tel: (514) 335-9274

DENMARK - DANEMARK
Munksgaard Export and Subscription Service
35, Nørre Søgade, DK-1370 København K
Tel. +45.1.12.85.70

FINLAND - FINLANDE
Akateeminen Kirjakauppa,
Keskuskatu 1, 00100 Helsinki 10   Tel. 0.12141

FRANCE
OCDE/OECD
Mail Orders/Commandes par correspondance :
2, rue André-Pascal,
75775 Paris Cedex 16
Tel. (1) 45.24.82.00
Bookshop/Librairie : 33, rue Octave-Feuillet
75016 Paris
Tel. (1) 45.24.81.67 or/ou (1) 45.24.81.81
Principal correspondant :
Librairie de l'Université,
12a, rue Nazareth,
13602 Aix-en-Provence   Tel. 42.26.18.08

GERMANY - ALLEMAGNE
OECD Publications and Information Centre,
4 Simrockstrasse,
5300 Bonn   Tel. (0228) 21.60.45

GREECE - GRÈCE
Librairie Kauffmann,
28, rue du Stade, 105 64 Athens   Tel. 322.21.60

HONG KONG
Government Information Services,
Publications (Sales) Office,
Beaconsfield House, 4/F.,
Queen's Road Central

ICELAND - ISLANDE
Snæbjörn Jónsson & Co., h.f.,
Hafnarstræti 4 & 9,
P.O.B. 1131 – Reykjavik
Tel. 13133/14281/11936

INDIA - INDE
Oxford Book and Stationery Co.,
Scindia House, New Delhi 1   Tel. 45896
17 Park St., Calcutta 700016   Tel. 240832

INDONESIA - INDONÉSIE
Pdii-Lipi, P.O. Box 3065/JKT.Jakarta
Tel. 583467

IRELAND - IRLANDE
TDC Publishers - Library Suppliers,
12 North Frederick Street, Dublin 1.
Tel. 744835-749677

ITALY - ITALIE
Libreria Commissionaria Sansoni,
Via Lamarmora 45, 50121 Firenze
Tel. 579751/584468
Via Bartolini 29, 20155 Milano   Tel. 365083
Sub-depositari :
Editrice e Libreria Herder,
Piazza Montecitorio 120, 00186 Roma
Tel. 6794628
Libreria Hœpli,
Via Hœpli 5, 20121 Milano   Tel. 865446
Libreria Scientifica
Dott. Lucio de Biasio "Aeiou"
Via Meravigli 16, 20123 Milano   Tel. 807679
Libreria Lattes,
Via Garibaldi 3, 10122 Torino   Tel. 519274
La diffusione delle edizioni OCSE è inoltre
assicurata dalle migliori librerie nelle città più
importanti.

JAPAN - JAPON
OECD Publications and Information Centre,
Landic Akasaka Bldg., 2-3-4 Akasaka,
Minato-ku, Tokyo 107   Tel. 586.2016

KOREA - CORÉE
Kyobo Book Centre Co. Ltd.
P.O.Box: Kwang Hwa Moon 1658,
Seoul   Tel. (REP) 730.78.91

LEBANON - LIBAN
Documenta Scientifica/Redico,
Edison Building, Bliss St.,
P.O.B. 5641, Beirut   Tel. 354429-344425

MALAYSIA - MALAISIE
University of Malaya Co-operative Bookshop
Ltd.,
P.O.Box 1127, Jalan Pantai Baru,
Kuala Lumpur   Tel. 577701/577072

NETHERLANDS - PAYS-BAS
Staatsuitgeverij
Chr. Plantijnstraat, 2 Postbus 20014
2500 EA S-Gravenhage   Tel. 070-789911
Voor bestellingen: Tel. 070-789880

NEW ZEALAND - NOUVELLE-ZÉLANDE
Government Printing Office Bookshops:
Auckland: Retail Bookshop, 25 Rutland Street,
Mail Orders, 85 Beach Road
Private Bag C.P.O.
Hamilton: Retail: Ward Street,
Mail Orders, P.O. Box 857
Wellington: Retail, Mulgrave Street, (Head
Office)
Cubacade World Trade Centre,
Mail Orders, Private Bag
Christchurch: Retail, 159 Hereford Street,
Mail Orders, Private Bag
Dunedin: Retail, Princes Street,
Mail Orders, P.O. Box 1104

NORWAY - NORVÈGE
Tanum-Karl Johan
Karl Johans gate 43, Oslo 1
PB 1177 Sentrum, 0107 Oslo 1Tel. (02) 42.93.10

PAKISTAN
Mirza Book Agency
65 Shahrah Quaid-E-Azam, Lahore 3 Tel. 66839

PORTUGAL
Livraria Portugal,
Rua do Carmo 70-74, 1117 Lisboa Codex.
Tel. 360582/3

SINGAPORE - SINGAPOUR
Information Publications Pte Ltd
Pei-Fu Industrial Building,
24 New Industrial Road No. 02-06
Singapore 1953   Tel. 2831786, 2831798

SPAIN - ESPAGNE
Mundi-Prensa Libros, S.A.,
Castelló 37, Apartado 1223, Madrid-28001
Tel. 431.33.99
Libreria Bosch, Ronda Universidad 11,
Barcelona 7   Tel. 317.53.08/317.53.58

SWEDEN - SUÈDE
AB CE Fritzes Kungl. Hovbokhandel,
Box 16356, S 103 27 STH,
Regeringsgatan 12,
DS Stockholm   Tel. (08) 23.89.00
Subscription Agency/Abonnements:
Wennergren-Williams AB,
Box 30004, S104 25 Stockholm.
Tel. (08)54.12.00

SWITZERLAND - SUISSE
OECD Publications and Information Centre,
4 Simrockstrasse,
5300 Bonn (Germany)   Tel. (0228) 21.60.45
Local Agent:
Librairie Payot,
6 rue Grenus, 1211 Genève 11
Tel. (022) 31.89.50

TAIWAN - FORMOSE
Good Faith Worldwide Int'l Co., Ltd.
9th floor, No. 118, Sec.2
Chung Hsiao E. Road
Taipei   Tel. 391.7396/391.7397

THAILAND - THAILANDE
Suksit Siam Co., Ltd.,
1715 Rama IV Rd.,
Samyam Bangkok 5   Tel. 2511630

TURKEY - TURQUIE
Kültur Yayinlari Is-Türk Ltd. Sti.
Atatürk Bulvari No: 191/Kat. 21
Kavaklidere/Ankara   Tel. 25.07.60
Dolmabahce Cad. No: 29
Besiktas/Istanbul   Tel. 160.71.88

UNITED KINGDOM - ROYAUME-UNI
H.M. Stationery Office,
Postal orders only:
P.O.B. 276, London SW8 5DT
Telephone orders: (01) 622.3316, or
Personal callers:
49 High Holborn, London WC1V 6HB
Branches at: Belfast, Birmingham,
Bristol, Edinburgh, Manchester

UNITED STATES - ÉTATS-UNIS
OECD Publications and Information Centre,
Suite 1207, 1750 Pennsylvania Ave., N.W.,
Washington, D.C. 20006 - 4582
Tel. (202) 724.1857

VENEZUELA
Libreria del Este,
Avda F. Miranda 52, Aptdo. 60337,
Edificio Galipan, Caracas 106
Tel. 32.23.01/33.26.04/31.58.38

YUGOSLAVIA - YOUGOSLAVIE
Jugoslovenska Knjiga, Knez Mihajlova 2,
P.O.B. 36, Beograd   Tel. 621.992

Orders and inquiries from countries where Sales
Agents have not yet been appointed should be sent
to:
OECD, Publications Service, Sales and
Distribution Division, 2, rue André-Pascal, 75775
PARIS CEDEX 16.

Les commandes provenant de pays où l'OCDE n'a
pas encore désigné de dépositaire peuvent être
adressées à :
OCDE, Service des Publications. Division des
Ventes et Distribution. 2. rue André-Pascal. 75775
PARIS CEDEX 16.

70431-01-1987

OECD PUBLICATIONS, 2, rue André-Pascal, 75775 PARIS CEDEX 16 - No. 43899 1987
PRINTED IN FRANCE
(97 87 02 1) ISBN 92-64-12921-9